Lecture Notes in Mathematics

Edited by A. Dold and B. Eckmann

675

Janos Galambos
Samuel Kotz

Characterizations of Probability Distributions

A Unified Approach with an Emphasis on Exponential and Related Models

Springer-Verlag
Berlin Heidelberg New York 1978

Authors
Janos Galambos
Samuel Kotz
Temple University
College of Liberal Arts
Department of Mathematics
Philadelphia, PA 19122/USA

AMS Subject Classifications (1970): 62 E 10, 62 H 05

ISBN 3-540-08933-0 Springer-Verlag Berlin Heidelberg New York
ISBN 0-387-08933-0 Springer-Verlag New York Heidelberg Berlin

Printing and binding: Beltz Offsetdruck, Hemsbach/Bergstr.
2141/3140-543210

PREFACE

The stimulus for this research project was supplied during the 1974 Calgary Conference on *Characterizations of Distributions* organized by G.P. Patil; both of us were active participants in this conference and could not help but notice the multitude of results on characterizations widely scattered in the literature.

Our aim in writing this monograph is twofold. One is to bring together the results in this useful and rapidly growing field in order to encourage further fruitful research in this area. We also have in mind a somewhat different objective, which in our opinion, is not less important and urgent at this time of rapid and unprecedented growth in periodical literature. The field of characterizations was independently developed in different branches of applied probability and pure mathematics which inevitably resulted in a certain amount of duplication of efforts. Moreover the terminology used in different sub-disciplines by various authors is not the same. Our second aim is therefore to unify the existing theory. We found that seemingly unrelated matheamtical theorems turned out to be part of the same general theory. We therefore tried to discover the unifying thread which passes through the majority of the papers written in the past one and a half decades in the field of characterizations of statistical distributions (in particular those dealing with the exponential distribution, both univariate and multivariate, and its monotonic transformations, which is the subject of this monograph). However, we should point out that there are still a number of as yet unsolved problems, for example, those stemming from Rossberg's results which require additional investigation.

This monograph contains no overlap with existing books on characterizations, since earlier books concentrated mainly on normality, which we do not discuss in our book. Moreover, as a rule, results dealing with exponential distributions which have appeared previously in other monographs or books are not reproduced here. This is true in particular for the material on this subject contained in Johnson and Kotz' (1972) *Multivariate Continuous Distributions* and the recent book by Galambos (1978) *The Asymptotic Theory of Extreme Order Statistics*, both of which were published by John Wiley and Sons.

To make the work self-contained, we have included the derivation of a number of results dealing with basic distribution theory, in particular those related to the distribution of order statistics. On the other hand, not all the theorems stated are supplemented with detailed proofs. In many instances, it was sufficient to present a comprehensive proof of the basic "unifying" theorem from which the subsequent results follow almost directly or require only minor modifications of the argument.

The level of mathematics was kept to a minimum, but the rigor of mathematical analysis was scrupulously maintained. Moreover, we have included extensive comments on the "practicality" of various mathematically equivalent characterizations. We thus feel that the monograph, which contains an up-to-date bibliography of over 250 items, will be useful and of interest to the general mathematical community, to probabilists and statisticians as well as to researchers and practitioners of industrial engineering and operations research and various scientists specializing in the natural and behavioral sciences, in particular those who are interested in the foundations and applications of probabilistic model building.

Philadelphia, Pa.
April, 1978.

Janos Galambos; Samuel Kotz

ACKNOWLEDGEMENTS

The authors acknowledge with thanks the financial support of the National Science Foundation during the period May 1976 - April 1977. It was during that period that the plans for this monograph were conceived and the broad outline of the project was formulated.

The authors are also indebted to their colleagues throughout the world - too numerous to mention individually - who generously supplied us with reprints and on occasion preprints of their work on characterizations of distributions and related topics which facilitated our task of compiling this monograph.

Thanks are also due to Ms. June Maxwell for her careful and accurate typing of the text.

TABLE OF CONTENTS

CHAPTER 1

PRELIMINARIES AND BASIC RESULTS

1.1. Introduction

The emergence of modern probability theory dates from the end of the nineteenth century and the early twentieth century, culminating with Kolmogorov's axioms in 1933.

Mathematical statistics were initiated even later: it was only in the early forties of the twentieth century that the basic concepts of this discipline crystalized. The flood of periodical and monographical literature in this area, which has persisted up until now at an ever accelerating rate, began after World War II.

At the same time, the theories of functions of real and complex variables, which serve as the foundations and basic tools of both probabilistic and mathematical-statistical arguments, reached their maturity at the end of the nineteenth century. In fact, the works of H. Lebesgue and A. Picard in the early part of the 20th century can be considered as the final development of these theories, at least as far as applications to nonabstract probability theory and mathematical statistics are concerned.

Probability theory and mathematical statistics are based on a number of specific concepts which are rather simple in meaning when the corresponding translation into pure mathematical terminology is made. Typical of these is the concept of the distribution function. It is merely a right continuous, nondecreasing and uniformly bounded function; moreover, the distribution of a sum of two independent random variables corresponds to the convolution of the two respective distribution functions. Some simple forms of conditional expectation correspond to integrals with predetermined upper or lower limits; the expectation of order statistics can be expressed as the integral of a polynomial of the distribution function, and so on. On the other hand, without this translation, an uncritical manipulation with concepts of probability theory and mathematical statistics may result, and has indeed resulted, in a multitude of seemingly unrelated theorems and corollaries. This, instead of clarifying the structure and properties of the basic notions of these disciplines, cause confusion and often completely obscure the relatively simple underlying natural interrelations.

There is perhaps no other branch of mathematical statistics in which this sad state of affairs is manifested so strongly as in the area dealing with characterizations of statistical distributions. Characterization theorems are located on the borderline between probability theory and mathematical statistics, and utilize numerous classical tools of mathematical analysis such as (advanced) theory of functions of complex variables, differential equations of different types, theory of series and, last but not least, the theory of functional equations.

It is the purpose of this monograph to show, by analyzing various characterizations of the exponential distribution and its natural transforms, the simplicity and clarity of characterizations of statistical distributions, when these are properly interpreted via elementary concepts of mathematical analysis and simple functional equations. We intend to demonstrate that almost all characterizations of the exponential distribution can be reduced to the analysis of one of five elementary properties enjoyed by the exponential function $G(x) = e^{-ax}$, $x \geq 0$. By doing so, we hope to reduce the further proliferation of artificial characterizations of the exponential distribution and its transforms. By "artificial" we mean characterizations which are expressed and often couched in terms of complicated concepts of mathematical statistics without the realization of its initial basic and completely equivalent pure and simple mathematical interpretation. This will perhaps allow the researchers in this area of statistics to concentrate on more meaningful and original problems which still await solution.

A major part of our monograph can be viewed as an imitation of a treatise written by a competent nineteenth century mathematician who is familiar with the basic concepts of modern probability theory and mathematical statistics but whose main emphasis is on the mathematical aspects of the problems.

Let us mention here that, to the best of our knowledge, there is no treatise in any language which would overlap with ours. The basic books on characterizations (the one by Lukàcs and Laha and the more comprehensive and more recent work by Kagan, Linnik and Rao) are geared almost entirely towards the normal distribution. In addition, their mathematical tools are much more advanced than ours.

Why has progress in solving characterization problems in mathematical statistics, in particular for the exponential distribution, been so erratic and inefficient? In our opinion, at least a partial answer to this question lies in the somewhat peculiar development of the theory of statistical distributions.

The development of probability theory has been overwhelmingly dominated by the (continuous) *normal* distribution since the early eighteenth century. Indeed, it started with DeMoivre's *Doctrine of Chances* (1713), then came Laplace's and Gauss' contributions in the nineteenth century; the beginnings of the twentieth century marked the culmination of the results related to the central limit theorem. The dominance of the normal distribution in statistics is perhaps even greater and only in the last two decades have we witnessed a mild retreat from, and rebellion against, its accepted universality. The development of Pearson's curves and Gram-Charlier and Edgeworth's series in this century are basically formal mathematical generalizations of the normal density via its characterizing properties.

It is therefore not surprising that the exponential distribution on its own appeared in statistical literature at a very late stage. As late as 1931, T. Kondo, in his paper in *Biometrika*, "Theory of Sampling Distribution of Standard Deviation," refers to the exponential distribution as Pearson's type X curve. (It should be pointed out that the study of the exponential distribution in actuarial, biological and engineering literature preceded its organized study in statistics by over 20 years; see Steffensen (1930), Teissier (1934) and Weibull (1939)). P.V. Sukhatme was probably the first, who in 1937 in the *Annals of Eugenics* suggested that the exponential distribution (which he called the chi-squared distribution with two degrees of freedom) could serve as an alternative to normal distributions "to the type of problems where the form of variation in the population is known and is far from normal." Sukhatme, in this same paper, determined the joint distribution of differences of order statistics from an exponential distribution, from which the following basic and important property follows. Namely, for the exponential distribution, the differences of order statistics are independent and exponential variates. This property was rediscovered a number of times by several authors including Malmquist (1950), Rényi (1953) and Epstein and Sobel (1953), whose papers had much influence on the development of the theory of exponential distributions in later years. We shall also use this result for characterization purposes in the present monograph.

One more look at the early history of probability and statistics can reveal that these theories were not ripe earlier for accepting the exponential distribution. The Poisson distribution (as an approximation to the binomial) appeared in Section 81 of the famous *Récherches sur la Probabilité des Jugements* (1837; or 1832 according to certain sources) by S.D. Poisson. But, as F.A. Haight pointed out in 1967, "Poisson then turned to other matters and did not refer to his discovery, either in the remainder of the book or in his subsequent publications." The Poisson distribution surfaced again in the literature some 60 years later in L. Bortkiewicz' monograph, *Das Gesetz der kleinen Zahlen* (1898) but did not win acceptance at this time either. It is noteworthy that Czuber's comprehensive monograph *Probability Theory* (1899) did not mention either the Poisson or the exponential distributions, in spite of Whitworth's *Choice and Chance* (1886) mentioning both and pointing out their strong relation. This relation also appeared in a physical context in Boltzmann's *Studien über das Gleichgewicht der lebendigen Kraft zwischen bewegten materiellen Punkten* (1868). But statistics had to wait until the 1950's for the general acceptance of the exponential distribution.

The groundbreaking paper was that of W. Weibull in 1951, in which he briefly investigated an extension of the exponential distribution which now bears his name. In 1952, D.J. Davis published a paper entitled "An analysis of some failure data" (*J. Amer. Statist. Assoc.*), in which he defines "the exponential theory of failures" and compares it with its normal counterpart in 13 applications ranging from bank statement errors to bus motor failures.

A serious study of the properties of the exponential distributions in statistical literature commenced with the publication in 1953 in the *J. Amer. Statist. Assoc.* of B. Epstein and M. Sobel's paper "Life testing" which started the ball rolling. A comprehensive bibliography of some 82 items on this subject up to 1969 can be found in the book by N.L. Johnson and S. Kotz: *Continuous Univariate Distri-*

butions I, pp. 228-32 (Wiley, New York, 1970).

This now explains why characterizations of the exponential distribution, or of its natural transforms, started so late. The basic lack of memory characterization of the exponential distribution is simply the logarithmic variant of the functional equation $f(x+y) = f(x)f(y)$, which is due to Cauchy (1821) and N. Lobačevskii (1829) (see, e.g., J. Aczél (1966)). (A complete solution of this equation for continuous and discontinuous cases was given by E.B. Wilson (1899) (the logarithmic variant) and by G. Hamel (1905).) This is discussed in detail in the first edition of Feller's book in 1950. It is conjectured by us--based on some indirect evidence--that the first *explicit* characterization of this distribution appeared sometime in the 19th century in literature devoted to the physical sciences, in particular, in connection with some earlier theories of the atom (since atoms exhibit the lack of memory property). Unfortunately, we have not as yet found the relevant references. The actual earliest explicit characterization of the exponential distribution known to us is via extreme value theory. Most prominently, the classical papers by Fréchet (1927) and by Fisher and Tippet (1928) should be mentioned (the exact statements and more details on this line of history are given in J. Galambos' book (1978)). These earlier contributions, however, did not have sufficient impact among probabilists or statisticians. The modern area of characterization theorems pertaining to the exponential distribution (and its monotone transforms) was originated by European mathematicians such as Fisz (1958) (characterizations via order statistics), basic and elaborate results of H.J. Rossberg (1960) (characterizations based on the range and ratios of order statistics) and Rényi (1956) (via Poisson processes and, more generally, renewal processes). Quite independently in the U.S.A., the first characterizations of the exponential distribution were due to Ghurey (1960) and Teicher (1961), who modified the available characterizations of the normal distribution to the exponential case (via the distribution of t-type statistics and by maximum likelihood property, respectively). This fact serves as an additional indication of the often undeserved prominence enjoyed by the normal distribution among a large group of statisticians. It is only recently that we detect a relaxation and even a drifting away from this direction. Here computer technology should receive credit for making possible this healthy widening of horizons in the field of statistical distribution theory and methodology.

As far as the normal distribution is concerned, the first modern characterization result is due perhaps to Pólya (1923) via independence of a variable and a linear combination of two variables. This was followed by Cramèr's (1936) result which states that if $X+Y$ is normal then so are X and Y, where X and Y are independent nondegenerate random variables. In the same year, Geary (1936) obtained that, within a large class of distributions, normality is characterized by the independence of the arithmetical mean and the variance of a sample. His result was refined by Lukàcs (1942), and put into final form by T. Kawata and H. Sakamoto (1949). The Pólya-Cramèr type of investigations were continued by M. Kac (1939) and S.N. Bernstein (1941), whose works can be considered as the foundations of a general theory. Lukàcs' method proved fruitful for another line of research in the theory of characterizations, which is well presented in the monograph of Lukàcs and Laha (1964). The more recent book by Kagan, Linnik and Rao (1973), on the other hand, gives a good collection of results of various kinds for the normal distribution. This is the main reason why we do not deal with normality in the present monograph.

In summary, the study of the exponential distribution within the framework of mathematical statistics started at a very late date and after numerous basic concepts of this discipline had been fully established. The theory of characterizations of the exponential distribution began even later. It was developed by statisticians who became used to modern statistical terminology and concepts without sufficient emphasis and attention to the underlying mathematical structure. As will be seen later, the difference in terminology usually means that one has to integrate by parts or has to use a substitution, but one arrives at the same integral equation. This neglect resulted in:

(a) a duplication or triplication of research efforts and unnecessary restatements of virtually identical results using different terminology. Laurent (1974) correctly points out that the fact that several characterization theorems are "rediscovered in applied statistical literature may be a consequence of compartmentalization and specialization."

(b) obscuring the simplicity of the mathematical properties associated with the exponential function and its monotone transformations.

Our aim is therefore to present an up-to-date account of the theory without the above stated misconceptions or oversights.

1.2. Preliminaries

For notational purposes, let us introduce the following basic concepts of probability theory.

We always denote the underlying probability space by (Ω,A,P), where Ω is the sample space, A is the set of events (σ-algebra of subsets of Ω) and P is the probability measure on A. Random variables will be denoted by capital letters, in most cases without any explicit reference to space Ω. Thus X is a random variable, which is a function on Ω and which is measurable with respect to A; i.e., for any real number x, the set of elements of Ω for which $X < x$ is a member of A. We again suppress $\omega\in\Omega$ in this notation and we abbreviate

$$\{\omega : X(\omega) < x, \omega \in \Omega\} = \{X < x\} .$$

For a given random variable X, the function

$$F(x) = P(X < x)$$

is called its *distribution function.* When necessary we emphasize X by adding a subscript such as $F_X(x)$.

With this definition, a distribution function $F(x)$ is nondecreasing, continuous from the left and $F(+\infty) = 1$, $F(-\infty) = 0$.

We call a distribution function $F(x)$ *degenerate* at the point a if

$$F(x) = \begin{cases} 1 & \text{if } x > a \\ 0 & \text{otherwise} . \end{cases}$$

If $F(x)$ has no point of discontinuity, we then refer to it as a *continuous* distribution function. To define discrete and absolutely continuous distribution functions, we introduce a concept for sequences of real numbers.

We call a sequence $u_1, u_2, \ldots$ of real numbers *simple* if $|u_j|$, $j \geq 1$, can be arranged in a nondecreasing order and if every finite interval (a,b) contains at most a finite number of the u_j.

Using this concept, $F(x)$ is called *absolutely continuous* if it is continuous and if it is differentiable for all x, except perhaps on a simple sequence. Finally, $F(x)$ is *discrete* if its set of discontinuities is a simple sequence and if $F(x)$ is parallel to the x-axis between points of discontinuity (in other words, $F(x)$ is a *step* function).

We shall use the same descriptive adjective for a random variable as for its distribution function. Thus we speak of degenerate, continuous absolutely continuous and discrete random variables which means that their distribution functions belong to the respective categories.

In view of the general properties of distribution functions, the following remark concerning these special classes of distributions is in order. Since $F(x)$ is always nondecreasing, its set of points of discontinuity is denumerable and it is differentiable for almost all x in the sense of Lebesgue. One, however, cannot exploit these properties in characterization theorems because the points of discontinuity can form a nonsimple sequence. Furthermore, a set of measure zero can be so large that no analytical advantages can be gained from knowing a property for all points except on such a set (although in some cases almost sure properties suffice, see, e.g., Chapter 2). In particular, if $f(x) = F'(x)$, defined for almost all x, then $F(x)$ can be expressed as an integral of $f(x)$, but very little can be gained from such an integral representation. On the other hand, if $F(x)$ is absolutely continuous and if $f(x) = F'(x)$ for all x, except perhaps on a simple sequence, then

$$F(x) = \int_{-\infty}^{x} f(t)dt$$

is a Riemann improper integral and this equation holds for all x. When we have an equation determining $F(x)$, then it may be an advantage that $F(x)$ be replaced by an integral for all x. Hence a solution among absolutely continuous distributions may easily be found, but it may be a difficult task to show that a given equation does not have any other solution than absolutely continuous ones. We shall see several cases when this difficulty arises.

The following formula based on integration by parts will be utilized on several occasions in the sequel.

Lemma 1.2.1. Let $X \geq 0$ be a random variable with a distribution function $F(x)$. Assume that $E(X^a)$ is finite for some $a > 0$. Then, for any $0 \leq T < +\infty$,

$$(1) \quad \int_T^{+\infty} x^a dF(x) = T^a[1-F(T)] + a \int_T^{+\infty} x^{a-1}[1-F(x)]dx \ .$$

In particular, if $E(X)$ is finite, then

$$(2) \quad \int_T^{+\infty} (x-T)dF(x) = \int_T^{+\infty} [1-F(x)]dx \ .$$

Proof: Let $T < N < +\infty$ be arbitrary. Then integrating parts yields

$$\int_T^N x^a dF(x) = N^a F(N) - T^a F(T) - a \int_T^N x^{a-1} F(x)dx$$
$$= T^a[1-F(T)] - N^a[1-F(N)] + a \int_T^N x^{a-1}[1-F(x)]dx \ .$$

Since, for any $T \leq N < +\infty$,

$$\int_T^N x^a dF(x) + \int_N^{+\infty} x^a dF(x) = \int_T^{+\infty} x^a dF(x)$$

and the integral on the right hand side is finite, the second term on the left hand side converges to zero as $N \to +\infty$. Hence, in view of the inequality

$$\int_N^{+\infty} x^a dF(x) \geq N^a \int_N^{+\infty} dF(x) = N^a[1-F(N)] \ ,$$

$N^a[1-F(N)] \to 0$ as $N \to +\infty$. Formula (1) thus follows. On the other hand, (2) is a special case of (1) with $a = 1$, hence the proof is completed. □

1.3. The classical case of the lack of memory property.

It is convenient to describe the classical case of the lack of memory property of a random variable or of its distribution in terms of life length of a hypothetical industrial device. Let X be the random life length of an item. Evidently, $X \geq 0$. We say that X, or the distribution function $F(x)$ of X, has the lack of memory property if, for all x, $z \geq 0$, for which $P(X \geq z) > 0$,

$$(3) \quad P(X \geq x+z \mid X \geq z) = P(X \geq x) \ .$$

If $P(X=0) \neq 1$, then for at least one $z_0 > 0$, $P(X \geq z_0) > 0$. But then a sequential argument yields from (3) that, for any integer $n \geq 1$, $P(X \geq nz_0) > 0$ and thus $P(X \geq z) > 0$ for any real number $z \geq 0$. Hence, we can rewrite (3) in the form

$$(3a) \quad P(X \geq x+z) = P(X \geq x)P(X \geq z) \ , \quad \text{all } x,z \geq 0$$

or equivalently

(3b) $1-F(x+z) = [1-F(x)][1-F(z)]$, all $x,z \geq 0$.

Although (3) and (3a,b) are equivalent only under the assumption that $F(x)$ is nondegenerate at zero, we can consider (3b) as the equation for the lack of memory, since it yields all the solutions of (3), including the degenerate one. The term "classical case" refers to the domain of (3b), namely, that we require it for all $x,z \geq 0$. In Chapter 2, we shall discuss extensions of this property in several directions and in Section 5 of this chapter, we shall analyze a statistical meaning of (3b) in the particular case $x = z$.

The solution of (3b) is now well known. We present this solution in

Theorem 1.3.1. There are only two solutions of the equation (3b) among distribution functions. Either $F(x)$ is degenerate at zero, or, with some contant $b > 0$, $F(x) = 1-e^{-bx}$, $x \geq 0$.

Remark. Obviously, the distribution function which is degenerate at zero is a solution of (3b). Hence we have to prove that if $F(x)$ is a solution of (3b) which is nondegenerate at zero then $F(x)$ is of the exponential form as stated in the theorem.

First Proof: We first present the classical method of solving (3b). Set $G(x) = 1-F(x)$. Then, by induction, (3b) implies that for any $x_j \geq 0$, $1 \leq j \leq n$,

(4) $G(x_1+x_2+\ldots+x_n) = G(x_1)G(x_2)\ldots G(x_n)$.

If we choose $x_j = x$ for all j, we obtain

(4a) $G(nx) = G^n(x)$

for any integral value of n. With $x = 1/n$ we now have

$$G(1) = G^n(\tfrac{1}{n}) \quad \text{or} \quad G(\tfrac{1}{n}) = G^{1/n}(1) .$$

From this and (4a) with $x = 1/s$, it follows that

(5) $G(\frac{n}{s}) = G^{n/s}(1)$.

Notice that $0 < G(1) < 1$. Indeed, if $G(1) = 0$ or $G(1) = 1$, (4a) would yield $G(y) \equiv 0$ or $G(y) \equiv 1$ which is not possible for $G(y) = 1-F(y)$ with $F(0) = 0$ and $F(x)$'s being nondegenerate at zero. Hence $b = -\log G(1)$ is finite and positive. We can therefore rewrite (5) as

(5a) $G(x) = e^{-bx}$ for any rational x.

Now let x be irrational. Let $\{x_1^{(n)}\}$ and $\{x_2^{(n)}\}$ be two sequences of rational numbers such that

$$x_1^{(n)} < x < x_2^{(n)} \quad \text{and} \quad \lim x_1^{(n)} = \lim x_2^{(n)} = x ,$$

where $n \to +\infty$. Since $G(x)$ is nonincreasing, (5a) yields

$$\exp(-bx_2^{(n)}) \leq G(x) \leq \exp(-bx_1^{(n)}) , \quad n \geq 1 .$$

Letting $n \to +\infty$ results in $G(x) = e^{-bx}$ for all $x \geq 0$, which was to be proved. □

Second Proof: Let us put again $G(x) = 1-F(x)$. Since $F(x)$ is not degenerate at zero, there is an $x_0 > 0$ such that $0 < G(x_0) < 1$ (otherwise, by (4a), $G(x)$ would be identically one or zero). In this case, however, $0 < G(x) < 1$ for any $x > 0$. Indeed, let n be such that $nx_0 \leq x < (n+1)x_0$. Then, since $G(x)$ is nonincreasing,

$$G[(n+1)x_0] \leq G(x) \leq G(nx_0) \ .$$

On the other hand, by (4a) both $G[(n+1)x_0]$ and $G(nx_0)$ are between zero and one.

Hence, on any finite interval both $G(x)$ and $1/G(x)$ are finite and different from one. Now let $b > 0$ be chosen in such a way that the function

$$(6) \qquad H(x) = G(x)e^{bx}$$

satisfies $H(x_0) = 1$. Then $H(x)$ satisfies

$$(7) \qquad H(x+z) = H(x)H(z)$$

and, for any $x \geq 0$,

$$(8) \qquad H(x+x_0) = H(x) \ .$$

Moreover, both $H(x)$ and $1/H(x)$ are finite on an arbitrary finite interval. Assume now that there is a $0 < x_1 < x_0$ such that $H(x_1) \neq 1$. We may assume that $H(x_1) > 1$, because otherwise we could argue with the function $1/H(x)$, which also satisfies the boundedness property as well as equations (7) and (8). Now, by (8), for any integer $N \geq 1$,

$$H(Nx_1+x_0) = H(Nx_1) \ ,$$

and thus by (7)

$$H(Nx_1+x_0) = H^N(x_1) \ .$$

Since $H(x_1) > 1$, this implies that, as $N \to +\infty$,

$$\lim H(Nx_1+x_0) = +\infty \ .$$

However, this limit leads to a contradiction. Indeed, by (8), if we write $Nx_1+x_0 = Mx_0+y$ with $0 \leq y < x_0$,

$$H(Nx_1+x_0) = H(y) ,$$

and thus $H(Nx_1+x_0)$ is bounded by the bound of $H(y)$ on the finite interval $[0,x_0)$, which was shown to be a finite number. Hence it is not possible that, for some $x_1 > 0$, $H(x_1) \neq 1$, and thus

$$G(x)e^{bx} = H(x) \equiv 1 \ ,$$

which proves the assertion. □

The first proof is the best known one and it reappears in the literature. It actually goes back to A.C. Cauchy (1821) and G. Darbaux (1875) (see, e.g., J. Aczél (1966)). The second proof is essentially due to G.S. Young (1958). We shall see further proofs in Section 5 of the present chapter (where several equivalent forms of (3b) are established) and several extensions are given in Chapter 2.

1.4. Motivations (Applications of characterization theorems in model building).

In applied statistics, one observes a random quantity X a number of times and, based on these observations, one would like to conclude facts about the distribution F(x) of X. The usual approach is to start with a family F of distributions and to select from this family a distribution F(x) which is the most acceptable one in a given sense. Unfortunately, in many cases, F simply consists of a single function which is dependent on one or several parameters, and the observations are used merely to approximate its parameters (estimation). The function thus obtained is chosen as F(x). Characterization theorems are the only methods which allow us to avoid the subjective choice of F and lead to the accurate F(x) through simple properties. To show how important the accuracy of the population distribution is, we borrow an example from the book of J. Galambos (1978).

Let Y be a standard lognormal variate and let $X = (Y^t-1)/t$, $t > 0$. It is well known (see, e.g., Johnson and Kotz (1970a)) that if t is small, then the distribution of X is close to the standard normal distribution. (Indeed, by Taylor's expansion

$$X = (Y^t-1)/t = (e^{t \log Y}-1)/t = \log Y + ut(\log Y)^2 ,$$

where $|u| \leq 1$. Hence, as $t \to 0$, the second term on the extreme right hand side tends to zero in probability. Thus, elementary probability theory shows that the distribution of X approaches that of log Y.) Let us therefore compare the values H(2.6) of the distribution function H(x) of the maximum of 50 observations on X under two different assumptions: X is normal and then $X = (Y^{0.1}-1)/0.1$. Easy calculations yield that

$H(2.6) = 0.59$ under normality

$H(2.6) = 0.78$ under lognormality .

The fact that goodness of fit tests would accept either one of the assumptions on the distribution of X, when the latter is the true form of X, can be seen from S. Kotz's (1973) analysis. It is evident that the decision should be different in the two cases, but classical statistics is simply unable to solve this controversy. An acceptable characterization theorem, however, can lead to a clear-cut distinction between such choices.

We describe here three applications of characterization theorems for illustration purposes. We wish to emphasize their simplicity and the nonmathematical nature of the underlying assumptions leading to the characterizations.

Insurance companies can collect accident records (history) of drivers. They label a driver "good" if the driver has the property that his probability of having an accident remains the same small number irrespective of the passage of time. In mathematical terms, this can be translated as follows. Let X be the random time period up to the first accident of the driver in question. Then

$$P(X \geq s + u | X \geq u) = P(X \geq s) .$$

This is the lack of memory property of the preceding section, which, as was seen, has a single nondegenerate solution in $P(X < x)$, namely the exponential distribution (we shall return to this point in the next section). Therefore, the insurance company is not subject to guessing in determining its risk for such a driver, but they can go ahead with a well defined model for setting the amount of the premium.

Let us now turn to another type of problem. Let a store service a community of n persons. These persons visit the store independently of each other and their actual times of entering the store have the same distribution. Therefore, the n individuals can be associated with n independent and identically distributed random variables, namely, with their random times X_j, $1 \leq j \leq n$, of entering the store. The store owner evidently observes the X_j in an increasing order. Let $X_{1:n} \leq X_{2:n} \leq \ldots \leq X_{n:n}$ be the successive arrivals at the store. Assume that the store owner observes that the interarrivals $X_{1:n}, X_{2:n}-X_{1:n}, \ldots, X_{n-1:n}-X_{n:n}$ are also independent. We shall see in Section 3.3 that this fact implies that the common distribution $F(x)$ of the X_j is again exponential. The emphasis here is that we have again arrived at a single possibility for $F(x)$. This means that the advice to all shops, or big department stores, depends on a single, well defined model, when a decision is to be made on the number of employees, availability of items, etc.

We formulate our third model in terms of telephones, but we shall see in Chapter 4, that this model is equivalent to several others (a fact which was not recognized earlier). Let calls arrive at a switchboard at independent and identically distributed random intervals. Each call reaches the intended person with probability p and the call does not go through with probability 1-p. If the intervals between the calls which go through are again independent and, apart from a scale factor, have the same distribution as the intervals between the actual calls, then the calls arrive according to a so-called Poisson process. The fact that a single model is obtained here greatly facilitates planning. For example, if a company receives sales orders through the telephone, the company can plan ahead with the number of telephone lines to be installed.

A single model in each of the preceding examples represents a characterization theorem. We hope that they will illustrate the advantage and importance of having a characterization result. On the other hand, our first example has pointed out the disadvantage of the lack of a characterization theorem for a specific problem. The succeeding sections contain a variety of characterization theorems. While we shall describe some other applied models as well, we do not intend to translate each theorem into a specific model. We would lose the advantages of mathematical abstraction by giving the impression that each theorem is presumably associated with a specific practical problem. This practice persists in the literature and this is one of the reasons that closely related results were rediscovered so many times without recognizing their apparent similarities.

1.5. Basic properties of the exponential distribution

Let X be a random variable with distribution function

$$(9)\quad F_X(x) = 1-e^{-b(x-a)}, \quad x \geq a,\ b > 0 .$$

Then X is called a (negative) exponential variate. Since Y = X + a has the distribution

$$(10)\quad F(x) = 1-e^{-bx}, \quad x \geq 0,\ b > 0 ,$$

many properties of X are equivalent to those of Y. A further simplification of the exponential distribution is possible by considering Z = bY. Its distribution is

$$(11)\quad F_Z(x) = 1-e^{-x}, \quad x \geq 0 ,$$

which is called the unit exponential distribution and Z a unit exponential variate.

The following properties of F(x) of (10) are immediate:

(P1) $(1-F(x))' = -b(1-F(x)), \quad x > 0,\ F(0+) = 0 ;$

(P2) $\int_z^{+\infty} (1-F(x))dx = \frac{1}{b}(1-F(z)) \quad \text{for all } z > 0,\ F(0+) = 0 ;$

(P3) $1-F(x+z) = (1-F(x))(1-F(z)) \quad \text{for all } x,z \geq 0 .$

Before going on with the discussion we wish to give the reason for including F(0+) = 0 in properties (P1) and (P2). Our aim with listing some basic properties of the exponential distribution is to arrive at characterization theorems. Now, if we drop F(0+) = 0 in (P1) and (P2) then evidently these properties remain to hold if 1-F(x) is replaced by c[1-F(x)] with arbitrary constant $c \geq 0$. Therefore, since $F(x) = 1-e^{-bx}$, b > 0, satisfies these properties, so do all functions

$$F_c(x) = \begin{cases} 1-ce^{-bx} & \text{if } x > 0 \\ 0 & \text{otherwise ,} \end{cases}$$

where $0 \leq c \leq 1$ and b>0. The above family of distributions has a jump at zero. That is, if X is a random variable with distribution function $F_c(x)$ then P(X=0) = 1-c. Consequently, (P1) and (P2) cannot lead to a characterization without some assumption on F(0+). The assumption F(0+) = 0 actually selects the exponential distribution (10) from the family $F_c(x)$. By this choice the degenerate distribution, that is represented by $F_c(x)$ with c = 0, is also excluded from (P1) and (P2), while (P3) is satisfied for the degenerate distribution at zero. Therefore, when we speak of the relation of (P1) and (P3), or (P2) and (P3), we limit our discussion to nondegenerate distributions.

Let us observe that (P3) can be written in the following equivalent form

(P3*) $P(Y \geq x+z | Y \geq z) = P(Y \geq x)$ for all $x,z \geq 0$.

Recall that the property (P3*) is called the *lack of memory*. While it is evident that (P1) and (P2) imply that (10) holds, and thus each of them is a characterization theorem, the equivalence of (P3) and (10) among nondegenerate distributions is much less evident. However, we have settled this equivalence in the preceding section. Let us draw the attention of the reader to the important fact that the domain for which a property holds is an essential part of that property. For example, if (P3), or its equivalent form (P3*), is assumed to be valid for some values of x and z only, then functions other than (10) may satisfy it. The most significant of these possibilities is the case when the equation of (P3) is required to hold for positive integers x and z only. Then its solution is the distribution function of a geometric variate U, whose distribution is given by the sequence $P(U=k) = p^k(1-p)$, $k = 0,1,2,\ldots, 0 < p < 1$. This fact immediately raises a question

(i) on how large a set $\{(x,z)\}$ must the equation of (P3) be valid to guarantee that its only solution be the function given in (10)?

We reformulate this question in a more general manner:

(ii) given any property which leads to the exponential distribution, how does this property relate to (P3*)?

We concentrate on (P3*) since it is a nonmathematical property and it is easy to decide in practice whether it can be assumed in a given situation. Namely, if we interpret Y related to (10) as the random life length of an item A, then (P3*) says that "A is not ageing," but its life is terminated by a sudden event. This interpretation is evident, since (P3*) can be translated as saying that if A has lived for at least z units of time, then it will live for an additional x units with the same probability as a new item similar to A. Therefore, if a property (Pj) characterizes the exponential distribution (10), then it is necessarily a new interpretation of the lack of memory property. The properties so far discussed are (Pj), $j = 1,2,3$, which are seen by the above indirect argument to be equivalent to (P3*). Let us, however, establish this equivalence by a direct mathematical proof. The equivalence of (P3) and (P3*) is of course trivial. We next establish the equivalence of (P1) and (P2). This again follows from a very simple and short argument. If we integrate (P1) from z to infinity, we obtain (P2). On the other hand, observing first that the left hand side of (P2) is continuous and thus so is the right hand side, we get that the two sides are differentiable. Differentiation now yields (P1), which shows that (P1) and (P2) are indeed equivalent. It now suffices to prove the equivalence of (P2) and (P3), say, in order to establish the equivalence of each of (Pj), $j = 1,2,3$ and (P3*) (recall that we limit our discussion to nondegenerate distributions).

We first show that, for a function satisfying (P3),

$$(12)\quad \int_0^{+\infty} (1-F(x))dx = 1/b$$

is finite. Indeed, if we apply (P3) n times with x = z, we get, for all integers $n \geq 1$ and all real numbers $z > 0$

$$(13a)\quad 1-F(nz) = (1-F(z))^n$$

or, for all integers $n \geq 1$ and all real numbers $x > 0$,

$$(13b)\quad 1-F(x) = (1-F(x/n))^n .$$

Thus, applying (13b) with x = n, we have

$$\int_0^{+\infty} (1-F(x))dx \leq \sum_{n=0}^{+\infty} (1-F(n)) = \sum_{n=0}^{+\infty} (1-F(1))^n = 1/F(1) ,$$

where $F(1) > 0$, since otherwise (13b) would imply that $F(x) = 0$ for all x, which contradicts the definition of a distribution function. If we now integrate (P3) with respect to x from zero to infinity, we get (P2) (the fact that $F(0+) = 0$ in (P3) is immediate since we assumed that F(x) is nondegenerate). The converse is easily seen as follows. If (P2) holds then so does (P1), as was seen earlier. But then the function $g(x) = e^{bx}(1-F(x))$ satisfies $g'(x) = 0$ and thus, with a suitable constant $B \neq 0$, $g(x)/B = 1$, for all $x > 0$. Consequently,

$$g(x+z)/B = \{g(x)/B\}\{g(z)/B\} \quad \text{for all } x,z > 0 ,$$

or, for all $x,z > 0$,

$$B(1-F(x+z)) = (1-F(x))(1-F(z)) .$$

Letting z tend to zero yields B = 1 and thus (P3) follows.

This rather brief discussion, when (13a) is properly reformulated, leads us to a new characterization and raises further problems. Namely, (13a) has the following important meaning. If a random variable Z has distribution F(x) and $Z_1, Z_2, \ldots, Z_n$ are independent observations on Z, then

$$Z_{1:n} = \min(Z_1, Z_2, \ldots, Z_n)$$

satisfies

$$P(Z_{1:n} \geq x) = (1-F(x))^n .$$

Thus, (13a) implies that, if F(x) satisfies the lack of memory property (P3*) then

(P4*) $Z_{1:n}$ is distributed as Z/n for all $n \geq 1$.

Let us show that (P4*) is in fact a characteristic property of the exponential distribution, provided that Z is nondegenerate. Evidently it is sufficient to prove that (P4*) implies (P3). In other words, we want to deduce the validity of (P3) from (13a,b). For this aim, we shall show that a function satisfying (13a,b) is continuous for all $x \geq 0$ and it satisfies (P3) for all rational $x,z > 0$. These two facts evidently imply the validity of (P3) for all $x,z > 0$. We first prove that $F(x)$ is continuous at $x = 0$. Indeed, since $F(x)$ is nondecreasing,(13a) implies that $F(0) = 0$. Since we have excluded from our investigation the possibility of $F(x)$ being degenerate at zero, (13a) also implies that $F(0+) = 0$, which yields our claim. Next, we deduce from (13a,b) that (P3) holds for rational values of $x,z > 0$. Putting $G(x) = 1-F(x)$, we have from (13a,b) that, for positive integers m and n,

$$G(m+n) = G(1)^{m+n} = G(1)^m G(1)^n = G(m)G(n)$$

and, for rational numbers $x = u/v$ and $z = p/q$,

$$G(x+z) = G(\frac{uq+vp}{qv}) = G(1/qv)^{uq+vp} = G(1/qv)^{uq} G(1/qv)^{vp}$$

$$= G(uq/qv)G(vp/qv) = G(x)G(z) .$$

We now show that $G(x)$ is continuous at an arbitrary $x = x_0$. In the above equation, let x and z be rational numbers such that $x < x_0 < x+z$. Let $x \to x_0$ and $z \to 0$. Because we have established that $G(x)$ is continuous at zero and that $G(0) = 1$, we obtain that $G(x_0+0) = G(x_0-0) = G(x_0)$, which proves the continuity of $G(x)$ at x_0.

Observe that since (P4*) is mathematically equivalent to (13a,b) the equivalence between (P4*) and (P3) is an answer to the question (i) posed earlier in this section. We do not discuss this question further here, since a complete answer to it is given in a separate section devoted to the lack of memory property. Here we only remark that the other solutions to this question involve only mathematical niceties.

Let us return to (P1) and (P2) and let us restate them in an applied scientist's language. For this purpose we shall use the following notation.

Definition. Let $F(x)$ be an absolutely continuous distribution function. Then its *hazard rate* $r(x)$ is defined as

$$r(x) = \frac{F'(x)}{1-F(x)} .$$

The concept of hazard rate plays a pivotal role in reliability theory, in insurance statistics (where it is termed the "force of mortality") and in biometrics. Evidently, (P1) is equivalent to

(P1*) the hazard rate is constant .

Turning to (P2), one can easily observe that it is related to the notion of conditional expectation. Let Z represent a life length and thus its distribution function $F(x)$ should satisfy $F(0) = 0$. Assume that $E(Z)$ is finite. Then, by definition,

$$E(Z-z|Z\geq z) = \int_{\Omega} (Z-z)dP(\cdot|Z\geq z) = \frac{1}{P(Z\geq z)} \int_{\{Z\geq z\}} (Z-z)dP$$

$$= \frac{1}{1-F(z)} \int_{z}^{+\infty} (x-z)dP(Z<x) .$$

Thus, by formula (2) of Lemma 1.2.1,

$$E(Z-z|Z\geq z) = \frac{\int_{z}^{+\infty} (1-F(x))dx}{1-F(z)} .$$

Therefore, (P2) can be restated as

(P2*) $E(Z-z|Z\geq z)$ is constant in $z, z \geq 0$ and $P(Z \leq 0) = 0$.

By translating (P1) and (P2) into (P1*) and (P2*), their seemingly purely mathematical content attains important practical significance. While an applied argument would hardly ever lead to (P2), its equivalent form (P2*) has a simple meaning in terms of applications. If an item has survived for z units, its remaining lifetime Z-z has exactly the same expected value as when it was new.

It is instructive to compare (P2*) and (P3*). From previous arguments it is clear that they are equivalent. However, if we examine their meaning, this equivalence is quite surprising, since logically (P2*) is a weaker assumption than (P3*), if one is not concerned with the fine mathematical points in the assumptions. Namely, in (P2), the integrability of 1-F(x) is an assumption, while in (P3), this property was deduced as a consequence of an initial assumption. These are, of course, implicit in their equivalent forms (P2*) and (P3*), which therefore should not be overlooked. Yet, when statistical methods are used to check a given assumption, (P2*) is indeed simpler than (P3*).

Let us emphasize here the fact that the four properties (P1*), (P2*), (P3*) and (P4*) represent four seemingly different applied problems and they have been used in the literature for model building. However, their interrelations have never been established. It is particularly pleasant to see that the property (P4*), which gives a characterization in terms of order statistics, is an integral part of the elementary theory of characterizations of the exponential distribution. The reader is also advised to keep the following fact in mind for the forthcoming chapters: since the four properties discussed are equivalent, an extension of one of these properties automatically gives an extension of the others as well. A summary of these four equivalent properties is presented in Table 1 on the next page.

In the preceding section we have given references for the classical solution of the lack of memory, which is our property (P3) or (P3*). Properties (P1) and (P2) and their solutions, can be found in any elementary textbook on calculus as the simplest possible form of a linear differential or integral equation. Strangely indeed, their reformulation in the forms of (P1*) and (P2*) induced much research and publication. (P2*) first appears as an exercise in the monograph of D.R. Cox

TABLE 1

formula number	mathematical formula	domain of validity	"applied" (practical) meaning	applied formula number
(P1)	$[1-F(x)]' = -b[1-F(x)]$	$x \ge 0$ $F(0+) = 0$	constant hazard rate; constant instantaneous failure rate; constant force of mortality	(P1*)
(P2)	$\int_z^{+\infty} [1-F(x)]dx = \frac{1}{b}[1-F(z)]$	$z > 0$ $F(0+) = 0$	constant residual life expectancy	(P2*)
(P3)	$1-F(x+z) = [1-F(x)][1-F(z)]$	$x,z \ge 0$ $F(x)$ nondegenerate	lack of memory; no ageing	(P3*)
(P4) [=(13a)]	$1-F(nz) = [1-F(z)]^n$	$n \ge 1$ integer $x > 0$ $F(x)$ nondegenerate	$nZ_{1:n}$ is distributed as the population	(P4*)

(1962, p. 128) and then in the papers of G. Guerrieri (1965), H.M. Cundy (1966) and H.E. Reinhardt (1968). Cundy mentions that he heard on a radio program that the expectation of life for the average sparrow is about 1.2 years, independently of the age of the bird. This induced his investigation, where the initial assumption is clearly (P2*) and it thus leads to the exponential distribution as the life distribution of sparrows. In the solution, Cundy assumes absolute continuity which was dropped by Reinhardt. The result is reobtained in many other publications. In several works, (P2*) is discussed in the following equivalent form: let $h(t)$ be a strictly increasing and nonnegative function and such that $h(A) = 0$ and $h(B) = +\infty$ for some $A < B$. Let T be a random variable and assume that

(14) $E\{h(T)|T>y\} = h(y) + h(b)$ for all $A \le y < B$,

where b is a constant. It is then deduced that $h(T)$ is exponential. The above property clearly coincides with (P2*), since, by assumptions, $\{T>y\} = \{h(T) > h(y)\}$ and thus with $Z = h(T)$ and $z = h(y)$, we have the following form of (14):

$$E(Z|Z>z) = z+h(b) \quad \text{for all } 0 \le z < +\infty .$$

This reformulation of (P2*) is due to M.A. Hamdan (1972) and it was followed by several others. (The main purpose of G.B. Swartz's (1975) short note is, e.g., to point out that (14) and (P2*) are equivalent.)

The fact that (P4*) characterizes the exponential distribution is a special case of the results by R.A. Fisher and L.H.C. Tippett (1928) and M. Fréchet (1927). It was also reobtained by M.M. Desu (1971), using a somewhat different approach.

We summarize properties (Pj) and (Pj*), $1 \le j \le 4$ in Table 1. In this table, we refer to (13a) as (P4), which is the mathematical equivalent of (P4*). The table is given on the preceding page.

Several extensions of the characterization of the exponential distribution based on (Pj) or (Pj*), $1 \le j \le 4$, are given in Chapters 2 and 3. They will provide a general answer to both of our questions (i) and (ii) posed in the beginning of the present section.

1.6. Monotonic transformations of the exponential distribution

Before proceeding with the discussion of the exponential distribution, we should point out that all of its properties can be translated into a property of an arbitrary continuous distribution. Indeed, if X is a random variable with continuous distribution function $F(x)$ then

(15) $Y = -\log F(X)$

or

(16) $W = -\log[1-F(X)]$

are unit exponential variates. Although not all properties of Y and W will be usable properties of X, and vice versa, most of them will. Therefore, a discussion of characterizations of the exponential distribution actually yields a family of characterization results. Let us write (15) and (16) in detail for a number of specific distributions which are widely used in applications. At this point, we rewrite only one property of the exponential distribution for each of the distributions listed. The obtained characterizing property is possibly the most indicative for the transformed distribution.

1. The uniform distribution. If $F(x) = x$ for $0 \le x \le 1$, then (15) implies that $Y = -\log X$ is unit exponential. The lack of memory property becomes

$$P(-\log X \ge x+z \mid -\log X \ge z) = P(-\log X \ge x), \text{ all } x,z > 0 .$$

With the substitution $e^{-x} = u$ and $e^{-z} = v$, we thus have directly from the lack of memory characterization that the property

$$P(X \le uv \mid X \le v) = P(X \le u) \ , \ 0 \le u, v \le 1$$

characterizes the uniform distribution. One could perhaps call it a lack of *multiplicative* memory as compared with the lack of *additive* memory for the exponential distribution.

2. The Weibull distribution. Let $a > 0$. Then the Weibull distribution is defined as

$$F(x) = 1 - \exp(-x^a) \ , \ x \ge 0 .$$

By (16), $W = X^a$ is a unit exponential variate. Because $a > 0$, we have immediately from (P4*) the following characteristic property of the Weibull distribution. If $X_{1:n}$ is the minimum of n independent observations on X then $n^{1/a}X_{1:n}$ is distributed as the population X.

3. The Pareto distribution. A random variable X with distribution function

$$F(x) = 1 - x^{-a}, \quad a > 0, \ x \ge 1,$$

is called a Pareto variate. The transformation (16) becomes here $W = a \log X$ and the lack of memory property states in this case that

$$P(X \ge uv \mid X \ge v) = P(X \ge u) \ , \ u,v \ge 1 .$$

We thus obtained that this property is a characterization of the Pareto distribution.

4. The extreme value distribution $\exp(-e^{-x})$. If the distribution of X is $F(x) = \exp(-e^{-x})$, then (15) yields that $Y = \exp(-X)$ is a unit exponential variate. We now transform (P4*) to obtain a characterization of the above distribution of X. Let $X_{n:n}$ be the maximum of n independent observations on X. Then X has the extreme value distribution F(x) if, and only if, $X_{n:n} - \log n$ is distributed as X for all $n \ge 1$.

5. The logistic distribution. We define the one parameter logistic distribution by the formula

$$F(x) = 1/(1+e^{-ax}) \ , \ a > 0, \ x \text{ real} \ .$$

By (16), $Y = \log(1+e^{-aX})$ is a unit exponential variate. Let us now determine the characteristic property of the logistic distribution via the property (P2) for Y. Since

$$P(Y > x) = P[X < -\frac{1}{a}\log(e^x-1)] \ , \ x > 0 \ ,$$

we have

$$\int_z^{+\infty} F[-\frac{1}{a}\log(e^x-1)]dx = \frac{1}{a} F[-\frac{1}{a}\log(e^z-1)]$$

as a characteristic property of the logistic distribution. This can of course be simplified by the substitution $u = (-1/a)\log(e^x-1)$. The equation above becomes

$$\int_{-\infty}^{t} \frac{F(u)}{1+e^{au}}du = a^{-2}F(t) \ .$$

Because order statistics will play an important role in several characterization theorems in the sequel, let us conclude this section with the following observation. If a double subscript as in $Z_{k:n}$ denotes the *k-th order statistic* of a sample of size n on a random variable Z, then the order statistics on the random variables of (15) and (16) are related by the formulas

(17) $Y_{k:n} = -\log F(X_{n-k+1:n})$

and

(18) $W_{k:n} = -\log[1-F(X_{k:n})]$.

These equations will be standard and frequently used for translating properties of the exponential distribution to those of the six distributions listed above. For example, differences of order statistics of the exponential distribution will be transformed into ratios of order statistics of the uniform distribution. Other transformations for the uniform as well as for other distributions, will be given in succeeding chapters.

CHAPTER 2

CHARACTERIZATIONS BASED ON TRUNCATED DISTRIBUTIONS

One can view the lack of memory property

(1) $P(X \geq x+z \mid X \geq z) = P(X \geq x)$, all $x,z \geq 0$,

where we assume that the distribution of X is nondegenerate, as a characterization of the exponential distribution based on a property on the truncated distribution $P(X \geq y \mid X \geq z)$, $y \geq z \geq 0$, of X. Another characterization based on a truncation property, namely, that

(2) $E(X \mid X \geq z) = z+c_1$, $z > 0$, $P(X \leq 0) = 0$,

where E(X) is finite and c_1 is constant, was shown in Section 5 of Chapter 1 to be equivalent to (1). The present chapter is devoted to several extensions of these characterizations of the exponential distribution.

Before we start with details, let us reemphasize that a characterization of a distribution obtained from (1) or (2) by a monotonic transformation should not be considered as an extension. Therefore, every characterization theorem for the exponential distribution is in fact a family of characterizations. For example, if we assume that

(2a) $E(X^2 \mid X \geq z) = z^2+c_2$, $z > 0$, $P(X \leq 0) = 0$,

where c_2 is a constant, then, with $h(x) = \sqrt{x}$ and $X^* = h(X)$, (2a) becomes (2), from which we know that X^* is a negative exponential variate. Consequently,

$$P(X<x) = P(X^* < \sqrt{x}) = 1-\exp(-b\sqrt{x}) , \quad x > 0 ,$$

is a Weibull variate. In this discussion the special form of h(x) above was, of course, immaterial; with an arbitrary strictly increasing h(x), (2) can be rewritten in an equivalent form like that of (2a). Similarly, the equation

(1a) $P(X^2 \geq x^2+z^2 \mid X \geq z) = P(X \geq x)$, all $x,z \geq 0$,

does not differ from (1). Because the sets $\{X \geq z\}$ and $\{X \geq x\}$ are not changed if X, x and z are replaced by their squares (since $x,z \geq 0$), (1a) becomes (1) with $X^2 = X_1$, $x^2 = x_1$ and $z^2 = z_1$. Hence, (1) implies that (1a) characterizes a Weibull variate again. Here, of course, X^2 can be replaced by X^α with arbitrary $\alpha > 0$. This fact was overlooked in the paper by Y.H. Wang (1976).

Let us now turn to some extensions of characterizing the exponential distribution by relaxing the conditions in (1) and (2). We state most theorems for the exponential distribution. But since they represent a family of characterizations, the reader is advised to restate each theorem in several equivalent forms. In particular, it is recommended to restate the results which follow for the distributions listed in Section 6 of Chapter 1.

2.1. Extended forms of the lack of memory property

We have seen in Section 5 of Chapter 1, that the condition "all $x,z \geq 0$" in (1) can be relaxed without enlarging the set of distributions satisfying equation (1). As a matter of fact, if we write (1) as

$$\text{(1b)} \quad 1-F(x+z) = [1-F(z)][1-F(x)] \ , \ x,z \geq 0 \ ,$$

where $F(x)$ is the distribution function of X, we obtain by induction

$$\text{(3)} \quad G(x_1+x_2+\ldots+x_n) = G(x_1)G(x_2)\ldots G(x_n) \ ,$$

where $x_j \geq 0$ and $G(x) = 1-F(x)$. Now, one of the results in Section 5 of Chapter 1 was that (3) characterizes the exponential distribution if (3) holds for all integers $n \geq 1$ and $x_1=x_2=\ldots=x_n = x \geq 0$. We now give several other extensions of the classical case.

We first prove a result presented in R. Fortet's monograph. (The English translation of this monograph appeared in 1977; see also the works of Jurkat (1965) and de Bruijn (1966).)

Theorem 2.1.1. Let $F(x)$ be a nondegenerate distribution function. Assume that $F(x)$ satisfies the equation of (1b) for almost all $x \geq 0$ and $z \geq 0$ with respect to Lebesgue measure. Then $F(x) = 1-e^{-bx}$, $x \geq 0$, where $b > 0$ is a constant.

Proof: Recall from Section 2 of Chapter 1 that any distribution function $F(x)$ is differentiable for almost all x in the sense of Lebesgue. Let $F'(x) = f(x)$ be the derivative of $F(x)$, which is therefore defined for almost all x.

Let us rearrange (1b) as

$$\text{(4)} \quad F(x+z) - F(z) = F(x)[1-F(z)]$$

and let us divide this equation by $x > 0$ to obtain

$$\text{(5)} \quad \frac{F(x+z)-F(z)}{x} = \frac{F(x)}{x}[1-F(z)] \ .$$

Now if $z \geq 0$ is such that $f(z)$ is defined, then the limit of the left hand side exists as $x \to 0+$. Hence

$$\lim_{x=0+} \frac{F(x)}{x} = b \geq 0$$

exists and it is finite. But then the limit of the left hand side of (5) exists for any $z \geq 0$ as $x \to 0+$. Denoting this limit--the right hand derivative--by $f^+(z)$, we get

(6) $f^+(z) = b[1-F(z)]$.

Furthermore, we have shown that $F(z)$ is continuous from the right for all $z \ge 0$. But any distribution function is continuous from the left and thus $F(z)$ satisfying (5) is continuous at every point. Hence, (4) can be rewritten as

$$F(y) - F(y-t) = F(t)[1-F(y-t)] \ , \ t > 0$$

from which, just as above, the existence of the left hand derivative $f^-(y)$ of $F(y)$ follows and, since $F(y)$ is continuous,

(7) $f^-(y) = b[1-F(y)]$, all $y \ge 0$.

By (6) and (7), $f(z)$ exists for all $z \ge 0$ and

$$f(z) = b[1-F(z)] \ .$$

The solution of this last equation is the claimed exponential function with $b > 0$, because $F(x)$ was assumed to be nondegenerate. The proof is completed. □

Let us modify the assumption of Theorem 2.1.1 as follows. Let Q be the uniquely determined measure on the Borel sets of the nonnegative real line which, for intervals $[c,d)$, $0 \le c < d$ is defined by

$$Q\{[c,d)\} = F(d) - F(c) \ , \ c \ge 0 \ .$$

Now, if we assume that B is a Borel set of nonnegative real numbers such that $Q(B) = 1$ and (1b) holds for all $x,z \in B$, then we are again dealing with the validity of (1b) for almost all x,z but now with respect to Q rather than with respect to Lebesgue measure. This change has a substantial effect on the conclusion. Namely, now not only the exponential function $F(x) = 1-e^{-bx}$, $x \ge 0$, $b > 0$, but other functions as well can satisfy (1b). For example, if X is a nonnegative integer-valued random variable with

$$P(X=k) = p^k(1-p) \ , \ 0 < p < 1, \ k = 0,1,2,\ldots \ ,$$

or, equivalently,

(8) $F(x) = P(X<x) = \sum_{0 \le k < x} p^k(1-p)$,

(a geometric variate), then $F(x)$ as given by (8) is a possible solution of (1b) if we choose $B = \{0,1,2,\ldots\}$. Indeed, B is a Borel set and $Q(B) = 1$ under (8). Furthermore, if $z \in B$, that is, z is a nonnegative integer, then

$$1-F(z) = \sum_{k=z}^{+\infty} p^k(1-p) = p^z(1-p)\sum_{k=0}^{+\infty} p^k = p^z \ ,$$

which evidently satisfies (1b) for all $x,z \in B$. The difficulty involved in this approach is that the set B is defined in terms of the unknown solution $F(x)$ of (1b) We do not discuss this problem in detail, but refer the reader to R.T. Durrett and S.G. Ghurye (1976). Our further discussion will be limited to those cases when the domain of the validity of the equation in (1b) does not depend on the unknown solution $F(x)$.

Theorem 2.1.2. Let $F(x)$ be a nondegenerate distribution function and put $G(x) = 1-F(x)$. Assume that (3) holds for two integral values of n, n_1 and n_2, say, such that $(\log n_1)/(\log n_2)$ is irrational, where in (3), $x_1=x_2=\ldots=x_n = x \geq 0$ is arbitrary. Then $F(x) = 1-e^{-bx}$, $x \geq 0$, with some $b > 0$.

This result is due to J. Sethuraman (1965). His result is actually more general than the statement above; the more general case will be discussed in subsequent chapters. Theorem 2.1.2 was later reobtained by B. Arnold (1971) (assuming $n_2 = n_1+1$), whose proof is different from the one presented below.

Proof: By assumption,

$$G^n(x) = G(nx) \quad \text{for all } x \geq 0 \quad \text{and for } n = n_1 \text{ or } n_2 .$$

Since this equation can also be written as $G(x/n) = G^{1/n}(x)$, we get by induction that, for any integer $N = n_1^s n_2^t$ with arbitrary integers s and t (both negative and positive),

$$(9) \quad G^N(x) = G(Nx) .$$

From the elements of mathematics we know that the set

$$(10) \quad u = \log N = s \log n_1 + t \log n_2$$

is dense on the real line as s and t go through the integers. Therefore, we can find two sequences $s = s(k)$ and $t = t(k)$ such that the corresponding $u = u(k) \to 0$. Let $u = u(k)$ be such a sequence. Then, if we put $g(x) = \log[-\log G(e^x)]$, $z = \log x$ and $u = \log N$, we get from (9)

$$g(u+z) - g(z) = u$$

or

$$(11) \quad \frac{g(u+z)-g(z)}{u} = 1 .$$

Hence, if $u \to 0$, we have $g'(z) = 1$ whenever the derivative exists (it does not follow from the last equation that the derivative exists because the sequence u is not arbitrary). Now, because $G(z)$ is monotonic, so is $g(z)$. In addition, the set u of the form of (10) is dense in R^1. Therefore, if $\{v\}$ is an arbitrary set such that $v \to 0$ then we can construct two sequences $u_1 < v \leq u_2$ such that $u_i \to 0$, u_i is of the form of (10) and $u_i/v \to 1$ for $i=1,2$. Consequently

$$\frac{g(u_1+z)-g(z)}{u_1} \frac{u_1}{v} \leq \frac{g(v+z)-g(z)}{v} \leq \frac{g(u_2+z)-g(z)}{u_2} \frac{u_2}{v} ,$$

which, by (11), reduce to

$$\frac{u_1}{v} \le \frac{g(v+z)-g(z)}{v} \le \frac{u_2}{v} .$$

Since the two extreme sides tend to one as $v \to 0$, $g'(z)$ exists and equals one. This now gives

$$G(e^z) = \exp(-e^{z+c}) , \text{ i.e., } G(x) = e^{-bx} , b > 0 ,$$

which was to be proved. □

By a similar argument, one can prove the following result of G. Marsaglia and A. Tubilla (1975).

Theorem 2.1.3. If a distribution function $F(x)$ satisfies (1b) for two values z_1 and z_2 of z such that $0 < z_1 < z_2$ and z_1/z_2 is irrational and for all values of $x \ge 0$, then $F(x)$ is exponential.

It may sound artificial to assume that z_1/z_2 is irrational. But without this assumption, the remaining conditions of Theorem 2.1.3 do not lead to a unique distribution $F(x)$. For example, if we assume that equation (1b) is satisfied for $z = 1$ and 2 and for all $x \ge 0$, then several distributions can be obtained as its solutions. Indeed, if $T(x)$ is an arbitrary distribution function with $T(0+) = 0$ then if we write $x > 0$ as $x = n+y$ where $n \ge 0$ is an integer and $0 \le y < 1$,

(12) $1-F(x) = p^n[1-T(y)] , \; p=1-F(1) ,$

satisfies (1b) for all integers $z \ge 0$ and all $x \ge 0$.

We can, of course, understand the mathematical reason for the need of z_1/z_2 to be assumed to be irrational. Just as in the case of Theorem 2.1.2, an essential step of the proof of Theorem 2.1.3 is the fact that the set

$$\{u: u = sz_1 + tz_2 , \; s,t \text{ integers}\}$$

is dense on the real line if z_1/z_2 is irrational, while this fails for a rational value of z_1/z_2.

In the discussion following the proof of Theorem 2.1.1 we pointed out that a modification of the assumptions of Theorem 2.1.1 can lead, among others, to a geometric variate. The geometric distribution

(13) $P(X=k) = p^k(1-p) , \; 0 < p < 1, \; k = 0,1,2,\ldots$

has a natural relation to the exponential in that the survival function, for integer x, can be represented as

$$P(X\ge x) = (1-p) \sum_{k=x}^{+\infty} p^k = p^x = e^{-bx} ,$$

where $b = \log(1/p) > 0$, and is therefore formally the same as the survival function of the exponential distribution. Therefore, an important fact about the distribution given by (12) is that it has no similarity to the exponential distribution.

The fact that the functional equation

(14) $f(x+z) = f(x)f(z)$ for some x and z ,

may have solutions which do not resemble the exponential function was first pointed out by M.V. Menon (1966). J. Aczèl (1975) gives a complete solution of (14) if x and z are assumed to belong to a finite segment of an arithmetic progression (i.e., to a finite lattice).

We now turn to another kind of extension of the classical lack of memory characterization. Let us write (1) in the form

(15) $P(X \geq x+z) = P(X \geq x)P(X \geq z)$.

The constants x and z can be considered as random variables with degenerate distributions. In other words, if Y and Z are random variables and if

(16) $P(X \geq Y+Z) = P(X \geq Y)P(X \geq Z)$

holds for Y and Z with degenerate distributions then X is necessarily exponential (Theorem 1.3.1). Following this approach, A. Obretenov (1970) gives a new proof of Theorem 1.3.1 using Laplace transforms. On the other hand, N. Krishnaji (1971) omits the assumption of Y's and Z's being degenerate. His main result is as follows:

Theorem 2.1.4. Let D_1 and D_2 be two families of random variables such that (i) each member of both D_1 and D_2 is independent of X, (ii) each member of D_1 is independent of each member of D_2 and (iii) for every interval $[a,b)$, $0 \leq a < b < +\infty$, there is a member V_i of D_i such that $P(a \leq V_i < b) > 0$, $i=1,2$. Assume that the distribution $F(x)$ of X is continuous for $x \geq 0$, $F(0+) = 0$ and satisfies

(17) $1-F(y+z) \geq [1-F(y)][1-F(z)]$ for all $y,z \geq 0$.

If (16) holds for all Y from D_1 and all Z from D_2, then $F(x)$ is exponential.

Proof: If equality always applies in (17), then the conclusion follows directly from Theorem 1.3.1 without any further assumptions. Therefore, let y_1 and z_1 be such that strict inequality applies in (17) with $y = y_1$ and $z = z_1$. Since $F(\cdot)$ is continuous by assumption, there are numbers $0 \leq a < b < +\infty$ and $0 \leq c < d < +\infty$ such that $a \leq y_1 < b$ and $c \leq z_1 < d$, and strict inequality applies in (17) for all $a \leq y \leq b$ and $c \leq z \leq d$. Let us fix a,b,c and d. Let Y and Z be members of D_1 and D_2, respectively, and such that $P(a \leq Y < b) > 0$ and $P(c \leq Z < d) > 0$. For this choice of Y and Z, (16) yields

(18) $$0 = \int_{-\infty}^{+\infty}\int_{-\infty}^{+\infty} [G(y+z) - G(y)G(z)]dP(Y<y)dP(Z<z)$$

$$\geq \int_a^b\int_c^d [G(y+z) - G(y)G(z)]dP(Y<y)dP(Z<z) ,$$

where, as usual, $G(x) = 1-F(x)$. Now, since $G(x)$ is continuous, the intervals $[a,b]$

and [c,d] are finite and strict inequality applies in (17) as y and z vary over [a,b] and [c,d], respectively, the theorem of Weierstrass in integral calculus implies that

$$G(y+z) - G(y)G(z) \geq \Delta > 0 \quad \text{for all } a \leq y \leq b,\ c \leq z \leq d .$$

Hence, by the choice of Y and Z, the right hand side of (18) is strictly positive which is a contradiction. Hence, strict inequality cannot apply in (17). The assertion of the theorem thus follows from Theorem 1.3.1. □

Notice that D_1 and D_2 may each contain a single random variable Y and Z, respectively. Then the assumptions on Y and Z are that they are independent of each other as well as of X and that for every $0 \leq a < b < +\infty$, $P(a \leq Y < b) > 0$ and $P(a \leq Z < b) > 0$. For this special case it would, at least, be interesting to relax the very restrictive assumptions on X: the validity of (17) and the continuity of F(x).

We conclude this section with an apparently new characterization of the logistic distribution which could be reduced to an extended form of the lack of memory. However, the transformed form would be of no interest, while the original characterization may have some practical value.

Theorem 2.1.5. Let the distribution function F(x) of the random variable X be continuous and symmetric about the origin. Then

$$F(x) = \frac{1}{1+e^{-\lambda x}}, \quad \lambda > 0 ,$$

if, and only if,

$$P(-x<X|X<x) = 1-e^{-\lambda x}, \quad \text{all } x \geq 0 ,$$

or if, and only if,

$$\frac{1-F(x+y)}{[1-F(x)][1-F(y)]} = \frac{F(x+y)}{F(x)F(y)} \quad \text{for all } x, y \geq 0 .$$

Proof: The logistic distribution $F(x) = (1+e^{-\lambda x})^{-1}$ is evidently continuous, symmetric about the origin, i.e., $F(-x) = 1-F(x)$ for all $x \geq 0$, and satisfies both assumptions of the theorem. Hence, only the converse needs proof.

Because F(x) is symmetric about the origin,

$$P(-x<X|X<x) = \frac{2F(x)-1}{F(x)} .$$

Since this conditional distribution was assumed to be exponential,

$$\frac{2F(x)-1}{F(x)} = 1-e^{-\lambda x} ,$$

the solution of which is evidently the claimed logistic distribution.

Now let us assume that the second property holds. By the assumption about symmetry,

$$G(x) = P(X<-x|X<x) = \frac{1-F(x)}{F(x)} ,$$

and thus the second assumption reduces to

$$G(x+y) = G(x)G(y) \quad , \text{ all } x,y \geq 0 \ .$$

In view of Theorem 1.3.1, its only continuous solution is $G(x) = e^{-\lambda x}$ with some $\lambda > 0$, $x \geq 0$. Hence, the first part of the theorem is applicable, from which it follows that $F(x)$ is a logistic distribution function. The proof is completed. □

The reader is advised to restate the theorem in the exponential form. That is, transform X by our rule of Chapter 1 to the variable

$$Y = \log(1+e^{-X})$$

and reformulate the assumptions in terms of Y. It then becomes clear that the theorem in the exponential version is too complicated to be of interest.

2.2. A stability theorem

As it was already indicated in Chapter 1, the subject matter of the theory of characterizations is to show that only one distribution (which may contain one or several parameters) can satisfy a set of specific properties. In most applications, however, statistical or mathematical properties can be verified only approximately. The question thus arises whether several different distributions can be the result of an investigation if a property is slightly modified. This area of research is called the stability theory of characterizations. In an abstract form, we can formulate the problem as follows.

Let A be a property which members of a family $\mathcal{F}$ of distributions may or may not possess. If there is only one member F of $\mathcal{F}$ that satisfies A then A leads to a characterization theorem (as an example, let A be the lack of memory property and $\mathcal{F}$ be the family of all nondegenerate distributions). Assume that B is another property which can be defined for members of $\mathcal{F}$. Furthermore, let the properties A and B be such that one can assign a precise meaning to the statement: "the distance $d(A,B)$" between A and B is smaller than $\varepsilon > 0$. Let $D(F_1,F_2)$ be a distance between any two members F_1 and F_2 of $\mathcal{F}$. The characteristic property A of F is called stable if, to any $\varepsilon > 0$, there exists a $\delta = \delta(\varepsilon)$ such that

$$\sup\{D(F,F_1): F \text{ possesses A and } F_1 \text{ possesses B}\} < \delta$$

whenever $d(A,B) < \varepsilon$.

Note that a stability theorem has two arbitrary distance concepts involved. Therefore, one can prove several stability theorems in connection with a single characterization theorem. While a complete unification is hardly possible, we wish to draw the attention of the reader to a recent attempt in this direction by V.M. Zolotarev (1976).

We now present one stability theorem in connection with the lack of memory property. This result is due to H.N. Hoang (1968). We introduce this theorem by going through our definition of stability step-by-step which can also serve to reduce the abstract nature of this definition.

Let F be the family of all continuous distributions $F(\cdot)$ with $F(0) = 0$. Let X be a random variable whose distribution belongs to F. Set

$$(19)\quad h(x,y) = P(X\ge x+y \mid X\ge y) - P(X\ge x) .$$

Let A be the property that $h(x,y) = 0$ for all $x \ge 0$ and $y \ge 0$, and B denote the property that $h(x,y)$ equals a prescribed function $h_1(x,y)$. We introduce

$$(20)\quad d(A,B) = \sup\{h_1(x,y): x \ge 0,\ y \ge 0\}$$

and

$$(21)\quad D(F_1,F_2) = \sup\{|F_1(x)-F_2(x)|: -\infty < x < +\infty\} ,$$

where F_1 and F_2 are arbitrary members of F (the latter is known in the literature as the Kolmogorov distance). We now establish Hoang's result.

Theorem 2.2.1. If the distribution function $F(x)$ of X belongs to F and if $F(x)$ has property B for which $d(A,B) < \varepsilon$, then there is a constant $b > 0$ such that

$$D(F(x),\ 1-e^{-bx}) < 2\sqrt{\varepsilon} .$$

In other words, if $F(x)$ almost satisfies the lack of memory then $F(x)$ is close to an exponential distribution in the sense of the Kolmogorov distance.

Proof: Put $G(x) = 1-F(x)$. Then (19) can be rewritten as

$$G(x+y) = G(y)[G(x)+h(x,y)] .$$

Thus we have

$$G(2x) = G^2(x) + G(x)h(x,x) ,$$

$$G(3x) = G(2x+x) = G^3(x) + G^2(x)h(x,x) + G(x)h(2x,x) ,$$

which can be continued to obtain

$$G(nx) = G^n(x) + G^{n-1}(x)h(x,x) + G^{n-2}(x)h(2x,x)+\dots+G(x)h[(n-1)x,x] .$$

Hence, using the assumption $d(A,B) < \varepsilon$, we have for any $x \ge 0$ for which $G(x) < 1$,

$$(22)\quad |G(nx) - G^n(x)| < \varepsilon \sum_{k=1}^{+\infty} G^k(x) = \frac{\varepsilon G(x)}{1-G(x)} \le \frac{\varepsilon}{1-G(x)} .$$

Now let $x_0 > 0$ and $b > 0$ be such that

$$(23)\quad 1-G(x_0) = F(x_0) = \sqrt{\varepsilon} = 1-\exp(-bx_0) .$$

Such an $x_0 > 0$ exists because $F(x)$ is continuous and $F(0) = 0$. Then, by (22), for all $n \geq 1$,

$$(24) \quad |G(nx_0) - e^{-nbx_0}| < \sqrt{\varepsilon}\ .$$

On the other hand, for an arbitrary $x > 0$ there is a unique integer $n \geq 1$ such that $(n-1)x_0 \leq x < nx_0$. Hence, from the monotonicity of $G(x)$ and by (24)

$$(25) \quad -\sqrt{\varepsilon} + e^{-nbx_0} < G(nx_0) \leq G(x) \leq G[(n-1)x_0] < e^{-(n-1)bx_0} + \sqrt{\varepsilon}$$

But, in view of (23),

$$0 \leq e^{-bx} - e^{-bnx_0} \leq e^{-b(n-1)x_0} - e^{-bnx_0} \leq 1-e^{-bx_0} = \sqrt{\varepsilon}$$

and similarly

$$0 \leq e^{-b(n-1)x_0} - e^{-bx} \leq \sqrt{\varepsilon}\ .$$

Consequently, (25) results in

$$|G(x) - e^{-bx}| < 2\sqrt{\varepsilon}$$

as claimed. □

The estimate in the conclusion can be improved. See T.A. Azlarov, *et al.* (1972).

2.3. Characterizations based on conditional expected values

Formula (2) expresses an elementary characterization of the exponential distribution. We have also mentioned that (2) can be restated in a number of equivalent forms by using monotonic transformations which lead to characterizations of other distributions without providing any new information. Our aim in this section is to obtain new characterization theorems by changing the assumptions in (2).

First notice that, by letting z tend to zero in (2), we get $c_1 = E(X)$. D.N. Shanbhag (1970) points out that if c_1 is specified in this way then the additional assumption $P(X\leq 0) = 0$ is superfluous. Namely,

$$E(X|X\geq 0) > E(X)/P(X\geq 0)\ ,$$

whenever $P(X<0) > 0$. Hence

$$E(X|X\geq z) = z + E(X)$$

fails as z goes to zero. On the other hand, if we drop $P(X\leq 0) = 0$ in (2), then it would not be a characteristic property of a single distribution anymore. As we pointed out in Section 1.5, a family $F_c(x)$ of distributions, each of which has a jump at zero, would satisfy (2). For these distributions, $c_1 = E(X)/[1-P(X=0)]$. Conversely, if we specify c_1 in (2) as

$$E(X|X\geq z) = z + kE(X),\ \text{all } z > 0,\ P(X\geq 0) = 1\ ,$$

where $k > 1$ is a given number, then this equation becomes a characteristic property of $F_c(x)$ with $c = 1/k$. That is, what we can conclude from this discussion is as follows. If we assume that $E(X|X\geq z)$ is of a special form (z + constant), then several distributions can be obtained for X ($F_c(x)$ with arbitrary c), but if we specify the unknowns as in the above equation (we give a meaning to the constant) then we arrive at a characterization theorem. With this general formulation we shall soon see that the linearity of $E(X|X\geq z)$ is not important. The simplest form of this fact, which is due to A.G. Laurent (initiated by him in the 1950's, but for convenience we refer to his published paper of 1974), is given below. A similar result was also obtained by M.N. Vartak (1974).

Theorem 2.3.1. Let $g(y) > 0$ be a decreasing and differentiable function for $0 \leq y < +\infty$. Assume further that $g'(y) \geq -1$ with $g'(0) \neq -1$ and such that

$$(26) \quad \int_0^A \frac{1+g'(y)}{g(y)}\, dy$$

is finite or infinite according as A is finite or infinite. If X is a nonnegative random variable with continuous distribution $F(x)$ and such that

$$(27) \quad E(X-y|X\geq y) = g(y)\ , \text{ all } y \geq 0\ ,$$

then the distribution $F(x)$ is uniquely determined. Specifically,

$$(28) \quad F(x) = 1-\exp\{-\int_0^x \frac{1+g'(y)}{g(y)}\, dy\}\ , \quad x \geq 0\ .$$

Proof: As we have seen in Section 1.5 (actually a consequence of Lemma 1.2.1),

$$E(X-y|X\geq y) = \frac{\int_y^{+\infty} [1-F(x)]dx}{1-F(y)}\ .$$

Hence, if we put $G(y) = 1-F(y)$, (27) becomes

$$g(y)G(y) = \int_y^{+\infty} G(x)dx\ .$$

Since by assumption $G(x)$ is continuous (continuity for $x > 0$ would follow from this last equation but not at $x = 0$), the right hand side is differentiable. Consequently, so is the left hand side. But $g'(y)$ exists by assumption and thus the existence of $G'(y)$ follows. We can now differentiate and we obtain

$$-\frac{G'(y)}{G(y)} = \frac{1+g'(y)}{g(y)}\ .$$

Integration and the assumption of continuity of $F(x)$ thus yield (28). The assumptions on $g(y)$ guarantee that $F(x)$ of (28) is indeed a distribution function. This completes the proof. □

When $g(y)$ is a given constant then we get back the negative exponential distribution. But since in most practical applications, a natural assumption is that

$g(y)$ is strictly decreasing, (28) provides the population distribution for a wide variety of problems.

Another modification of (2) can be that instead of considering the conditional expectation $E(X|X\geq z)$, we replace it by $E[h(X)|X\geq z]$ or $E[h(X-z)|X\geq z]$, where $h(\cdot)$ is a given function. In considering the equation

(29) $E[h(X)|X\geq z] = g(z)\ ,\ z \geq 0,\ P(X\geq 0) = 1\ ,$

however, we have to be careful not to repeat (2) in disguised but equivalent form, thus getting trivialities (see the discussion following (2a)). The other equation of interest which we mentioned above is

(30) $E[h(X-z)|X\geq z] = g(z)\ ,\ z \geq 0\ ,$

where $h(\cdot)$ and $g(\cdot)$ are given functions. The function $g(z)$, of course, has to be such that the resulting solution, if any, should be a distribution function. Further extension is possible by requiring the validity of (29) and (30) on an interval rather than on the whole positive real line.

All solutions of (29), known to the present authors, start with the assumption of $h(\cdot)$ being strictly monotonic. This case, however, can easily be reduced to the linear case (equation (27)), since $\{X\geq z\} = \{h(X) \geq h(z)\}$ for strictly increasing $h(\cdot)$. Therefore, this case does not present any new mathematical problem. On the other hand, (29) even with monotonic $h(\cdot)$ may give a new and meaningful characterization of distributions for which $E(X)$ is infinite. For example, if the distribution function of X is $F(x) = 1-1/x,\ x \geq 1$, then $E(X) = +\infty$ but

(31) $E(\sqrt{X}|X\geq z) = 2\sqrt{z}\ ,\ z \geq 1,\ P(X\geq 1) = 1\ ,$

uniquely characterizes the distribution $F(x) = 1-1/x,\ x \geq 1$, among all continuous distribution functions. Indeed, if we put $Y = \sqrt{X}$ and $u = \sqrt{z}$, then (31) reduces to

(32) $E(Y|Y\geq u) = 2u\ ,\ u \geq 1,\ P(Y\geq 1) = 1\ ,$

which is of the form (27). Now if we assume that X has a continuous distribution function then so does Y. Hence, all conditions of Theorem 2.3.1 are satisfied, with $g(u) = u$ (we apply Theorem 2.3.1 to $Y-1$ to make (32) applicable for all $u \geq 0$). Thus (28) yields

$$P(Y<u) = 1-\exp\{-\int_1^u \frac{2}{y}\,dy\} = 1 - \frac{1}{u^2}\ ,\quad u \geq 1\ .$$

Hence

$$F(x) = P(X<x) = P(Y^2<x) = P(Y<\sqrt{x}) = 1 - \frac{1}{x}\ ,\ x \geq 1\ ,$$

as claimed.

In connection with (29), for various choices of $g(\cdot)$, see the papers by I.I. Kotlarski (1972), D.N. Shanbhag and M. Bhaskara Rao (1975), R.C. Gupta (1975) and A.C. Dallas (1976a).

To show that (30) has a unique solution for certain given $g(z)$ is far more difficult. One cannot appeal here to inverse functions to reduce it to the linear case even if $h(\cdot)$ is assumed monotonic. Hence, a direct solution is required. Even the simplest case $h(u) = u^k$ for some $k \geq 2$ and $g(z) = E(X^k)$ (a constant) is significantly more difficult than the linear case given in Theorem 2.3.1. The difficulty lies in the fact that when we reduce (30) to a differential equation, then its order is k and its general solution is thus a linear combination of k independent solutions. From this, one has to deduce that only one of these linear combinations can be a distribution function. This last step can, of course, be avoided by additional assumptions, see H.N. Nagaraja (1975). The case $k = 2$ was treated by several authors (A.G. Laurent (1972 and 1974), T.A. Azlarov, *et al.* (1972) and A.C. Dallas (1975)). The case of arbitrary $k \geq 2$ was settled by O.M. Sahobov and A.A. Geshev (1974). A general theorem, which is applicable to a general class of functions $h(\cdot)$, was recently obtained by L.B. Klebanov (1977). Klebanov uses deep analytic tools. Because his paper has not yet appeared at the time of this writing, we do not give details. However, we present Sahobov and Geshev's result below.

Theorem 2.3.2. Let $X \geq 0$ be a random variable with distribution function $F(x)$. Assume that $E(X^k)$ is finite where $k \geq 2$ is a given integer. If

$$(33)\quad E[(X-z)^k | X \geq z] = E(X^k) \quad \text{for all } z \geq 0 ,$$

then $F(x) = 1-e^{-bx}$, $x \geq 0$, and $b > 0$.

Proof: Let $m = E(X^k)$. Since

$$E[(X-z)^k | X \geq z] = \int_z^{+\infty} (y-z)^k dF(y)/[1-F(z)] ,$$

an argument similar to the one applied in Lemma 1.2.1 and (32) yields

$$(34)\quad \int_z^{+\infty} (y-z)^{k-1} G(y)dy = \frac{m}{k} G(z) \quad , \ z \geq 0 ,$$

where $G(u) = 1-F(u)$. If we denote the left hand side of (34) by $H(z)$, then repeated differentiation in (34) gives

$$(35)\quad \frac{(-1)^k m}{k!} H^{(k)}(z) = H(z) .$$

It is well known from the elementary theory of linear differential equations that the general solution of (35) is

$$(36)\quad H(z) = \sum_{j=1}^{k} c_j e^{b_j z} ,$$

where the c_j are arbitrary constants and b_j, $1 \leq j \leq k$, are the k (complex) solutions of the equation

$$\frac{(-1)^k m}{k!} t^k = 1 .$$

Let its negative solution be denoted by $b_1 = -b$, $b > 0$ (evidently, b is the positive real root of $(k!)/m$). Because, by definition and by (34), $H(z) = (m/k)G(z)$, $H(z) \geq 0$, decreases and $H(z) \to 0$ as $z \to +\infty$. We shall show that these properties can be satisfied only if

$$(37) \quad \sum_{j=2}^{k} c_j e^{b_j z} = 0$$

identically. First let $k = 3$. Considering the conjugate complex roots, we have $\mathrm{Re}b_2 = \mathrm{Re}b_3 = r > 0$ and $r < b$. Then $|\exp(b_j z)| \to +\infty$ as $z \to +\infty$ for $j = 2,3$. Hence, $H(z) \geq 0$ and $H(z) \to 0$ can be satisfied only if $c_2 = c_3 = 0$. If $k = 4$, then $\mathrm{Re}b_2 = \mathrm{Re}b_3 = 0$ and $b_4 = b$. In a similar manner as above, we easily verify that $c_4 = 0$. On the other hand, since

$$c_2 e^{b_2 z} + c_3 e^{b_3 z}$$

has to be real, $H(z)$ cannot be decreasing for large z unless $c_2 = c_3$. But then the above expression is identically zero. Now the general case can be handled similarly. If one combines two roots b_j and b_{j+1}, which are complex conjugates of each other, step by step one indeed arrives at the validity of (37). Hence

$$H(z) = c_1 e^{-bz} , \quad b > 0 ,$$

from which

$$F(z) = 1-G(z) = 1 - \frac{kc_1}{m} e^{-bz} , \quad b > 0, \ z \geq 0 .$$

But since $m = E(X^k)$, (34) with $z = 0$ and Lemma 1.2.1 imply that $G(0+) = 1$ or $F(0+) = 0$. Consequently, $(kc_1)/m=1$, which completes the proof. □

Let us return to the linear case. K-M. Chong (1977) points out the following interesting property of the exponential distribution.

Theorem 2.3.3. Let $X \geq 0$ be a random variable with finite expectation. Let $(X-u)^+ = \max(X-u,0)$. If the distribution of X is continuous at zero, then X has a negative exponential distribution if, and only if,

$$(38) \quad E[(X-s)^+]E[(X-t)^+] = mE[(X-s-t)^+] , \quad \text{all } s,t \geq 0 ,$$

with some constant $m > 0$.

Proof: Because $X \geq 0$, setting $t = 0$ in (38) yields $m = E(X)$. If we put $g(u) = m^{-1}E[(X-u)^+]$, then the assumption leads to the equation

$$g(s)g(t) = g(s+t) , \quad \text{all } s,t \geq 0 .$$

Since $g(u)$ is evidently monotonic, Theorem 1.3.1 implies

$$E[(X-u)^+] = me^{-bu} \quad , \; u \geq 0$$

for some constant $b > 0$. The last equation becomes

$$\int_u^{+\infty} G(t)dt = me^{-bu} \quad , \; u \geq 0,$$

where $G(t) = P(X \geq t)$. The reader can now easily conclude the proof by differentiation and by utilizing the assumption that the distribution of X does not have a jump at zero. □

From the proof, it is clear that the theorem can be modified in the same manner as the lack of memory characterizations were modified in the beginning of this Chapter. For example, if "all $s,t \geq 0$" is changed to "all $s \geq 0$ and two values of t, t_1 and t_2, say, for which t_1/t_2 is irrational," the conclusion remains to hold. Other variants may be deduced from previous discussions. The assumption of continuity at zero cannot be dropped, however, because a jump of the distribution of X at zero has no effect on the other conditions.

T.A. Azlarov, *et al.* (1972) obtained stability theorems related to most of the characterizations discussed in the present chapter. Several of their theorems, however, were previously obtained by T.A. Azlarov (1972) in a somewhat weaker form. The recent survey by E. Lukacs (1977) gives a good account of the pre-1976 results on stability.

CHAPTER 3

CHARACTERIZATIONS BY PROPERTIES OF ORDER STATISTICS

3.1. Introduction

Let $X_1, X_2, \ldots, X_n$ be independent and identically distributed random variables with common distribution function F(x). Let us rearrange the X's in increasing order

$$X_{1:n} \le X_{2:n} \le \ldots \le X_{n:n} .$$

We call $X_{r:n}$ the r-th order statistic. If two values X_j and X_t are equal for $j \neq t$, then we do not stipulate which variable would precede the other one. If F(x) is continuous then $P(X_j = X_t) = 0$ for any $j \neq t$; therefore in this case the order statistics are distinct with probability one. Since our aim is to obtain characterization theorems for the exponential distribution and for its strictly monotonic transformations, we shall assume in most cases that F(x) is continuous. Therefore, the present chapter does not cover the literature of discrete distributions. The reader can find a comprehensive bibliography and a detailed exposition (with proofs and with several new results) for discrete distributions in J. Galambos (1975b).

Let us collect a few basic formulas from the theory of order statistics which will be needed in the present chapter.

By the basic assumption of the X_j's being independent and identically distributed,

$$(1) \quad P(X_{n:n} < x) = F^n(x)$$

and

$$(2) \quad P(X_{1:n} \ge x) = [1-F(x)]^n .$$

In general, the event $\{X_{r:n} < x\}$ means that at least r of the events $\{X_j < x\}$ occur. Hence, applying the binomial distribution with parameters n and F(x),

$$(3) \quad F_{r:n}(x) = P(X_{r:n} < x) = \sum_{j=r}^{n} \binom{n}{j} F^j(x)[1-F(x)]^{n-j} = r\binom{n}{r} \int_0^{F(x)} t^{r-1}(1-t)^{n-r} dt .$$

Formulas (1) and (2) are, of course, special cases of (3).

If the density f(x) of X_j exists, then (3) implies that so does the density $f_{r:n}(x)$ of $X_{r:n}$ for each r and

$$(4) \quad f_{r:n}(x) = r\binom{n}{r} F^{r-1}(x)[1-F(x)]^{n-r} f(x) .$$

Let us give a more direct proof of formula (4), which has the advantage of being applicable to the case of the joint density of several order statistics. Let

$\Delta x > 0$ and let $F(x)$ be absolutely continuous. Then the probability that at least two of the X_j fall into the interval $(x,x+\Delta x)$ is $O[(\Delta x)^2]$ as $\Delta x \to 0$. Hence $P(x \le X_{r:n} < x+\Delta x)$ can be calculated as follows. The event $\{x \le X_{r:n} < x+\Delta x\}$ approximately means that r of the X_j are smaller than $x+\Delta x$, one of which is between x and $x+\Delta x$ and the remaining r-1 are smaller than x. The rest are larger than $x+\Delta x$. Therefore

$$P(x \le X_{r:n} < x+ \Delta x) \approx r\binom{n}{r}F^{r-1}(x)[1-F(x)]^{n-r}[F(x+\Delta x)-F(x)] ,$$

the error being $O[(\Delta x)^2]$. If we divide by Δx and let $\Delta x \to 0$ we get an exact formula for $f_{r:n}(x)$, which is indeed the same as (4). This same method yields the following result. Let $1 \le r_1 < r_2 < \ldots < r_k \le n$ be integers. Let $F(x)$ be absolutely continuous. Then the joint density of the vector $(X_{r_1:n}, X_{r_2:n}, \ldots, X_{r_k:n})$ exists and equals

$$\text{(5)} \quad f_k(\underset{\sim}{x},\underset{\sim}{r}) = n![\prod_{j=1}^{k} f(x_j)] \prod_{j=0}^{k} \frac{[F(x_{j+1})-F(x_j)]^{r_{j+1}-r_j-1}}{(r_{j+1}-r_j-1)!} ,$$

whenever $x_1 \le x_2 \le \ldots \le x_k$ and zero otherwise, where $\underset{\sim}{x} = (x_1,x_2,\ldots,x_k)$ and $\underset{\sim}{r} = (r_1,r_2,\ldots,r_k)$, $x_0 = -\infty$, $x_{k+1} = +\infty$, $r_0 = 0$ and $r_{k+1} = n$.

For the exponential distribution $F(x) = 1-e^{-bx}$, $x \ge 0$, a direct substitution into (5) yields the following result, which was first observed by P.V. Sukhatme (1937).

Theorem 3.1.1. Let $F(x) = 1-e^{-bx}$, $x \ge 0$. Then the differences

$$\text{(6)} \quad d_{r:n} = X_{r+1:n} - X_{r:n} , \quad r \ge 0, \quad X_{0:n} = 0 ,$$

are independent exponential variables with

$$P(d_{r:n} < x) = 1-e^{-b(n-r)x} , \quad x \ge 0.$$

Another substitution into (5), namely that of $F(x) = x$, $0 \le x \le 1$, leads to a result for the uniform distribution.

Theorem 3.1.2. Let $F(x) = x$ for $0 \le x \le 1$. Then the differences $d_{r:n}$ of (6) are identically distributed with common distribution function $F_1(x) = 1-(1-x)^n$.

Notice that the differences $d_{r:n}$ in Theorem 3.1.2 are not independent. In fact, we shall show that the independence of the sequence $d_{r:n}$ is a characteristic property of the exponential distribution. However, formulas (17) and (18) of Section 1.6 can be applied to transform this property into one for arbitrary continuous distributions. For example, we immediately get from Theorem 3.1.1, that the ratios $X_{r+1:n}/X_{r:n}$ are independent for the uniform distribution $F(x) = x$, $0 \le x \le 1$.

We record yet one more property of order statistics.

Let $F(x)$ be continuous. Then, as remarked in Section 1.6, the transformed order statistics

$$W_{k:n} = -\log[1-F(X_{k:n})]$$

are order statistics from a unit exponential distribution. Hence, by Theorem 3.1.1, $W_{k:n}$ can be represented as a sum of independent random variables. From this representation it is immediate that $W_{k:n}$, $k \geq 1$, and thus $X_{k:n}$, $k \geq 1$, as well, form a Markov chain. The explicit form of the representation of $W_{k:n}$ as the sum mentioned above also implies the following important result.

Theorem 3.1.3. Let $F(x)$ be continuous. Then the conditional distribution $P(X_{k:n}<x|X_{r:n}=y)$, $k > r$, is the same as the distribution of the $(k-r)$-th order statistic in a sample of size $n-r$ from a distribution

$$F^*(x) = \begin{cases} \dfrac{F(x)-F(y)}{1-F(y)} & \text{if } x \geq y \\ 0 & \text{otherwise .} \end{cases}$$

The three theorems proved here will be basic to the characterization theorems of the succeeding sections. We have seen in Chapter 1, that certain characterization theorems based on $X_{1:n}$ are strongly related to the lack of memory property and to several of its equivalent forms. This fact links closely the content of the present chapter with the preceding ones.

3.2. Characterizations by distributional properties

One of the basic characterizations of the exponential distribution in Section 1.5 was property (P4*) which states that among the nondegenerate distributions only the exponential has the property that $nX_{1:n}$, for all $n \geq 1$, is distributed as the population. We have extended this characterization in Theorem 2.1.2 by relaxing the condition "for all $n \geq 1$" to "for two values n_1 and n_2 of n such that $(\log n_1)/(\log n_2)$ is irrational." This theorem can further be extended as follows. (This result is also due to J. Sethuraman (1965).)

Theorem 3.2.1. Let $F(x)$ be a nondegenerate distribution function. Assume that there are two sequences A_n and $B_n > 0$ of numbers such that $(X_{1:n}-A_n)/B_n$ is distributed as the population for two values n_1 and n_2 of n such that $B_{n_2} \neq 1$ and $(\log B_{n_1})/\log B_{n_2}$ is irrational. Then there are constants, a, b > 0 and c > 0 such that $F(a+bx)$ is either

$$L_{1,c}(x) = \begin{cases} 1-\exp[-(-x)^{-c}] & \text{if } x < 0 \\ 1 & \text{if } x \geq 0 \text{ ,} \end{cases}$$

or

$$L_{2,c}(x) = \begin{cases} 1-\exp(-x^{c}) & \text{if } x > 0 \\ 0 & \text{if } x \leq 0 \text{ .} \end{cases}$$

On the other hand, if $B_{n_2} = 1$ and A_{n_1}/A_{n_2} is irrational then $F(a+bx) = 1-\exp(-e^{x})$

with some constants a and b > 0.

We do not present here the proof of this theorem because it is practically the same as that of Theorem 2.1.2 after establishing that if $B_{n_2} > 1$ then so is B_{n_1} and $A_{n_1}/(1-B_{n_1}) = A_{n_2}/(1-B_{n_2})$.

If we now specify that $A_n = 0$ and $B_n = 1/n$ for $n = n_1$ and n_2, then to obtain Theorem 2.1.2 from Theorem 3.2.1 it is only necessary to check which of the three distributions can satisfy the assumptions. We thus get that, for $A_n = 0$, $B_n = 1/n$, the resulting distribution is $F(x) = L_{2,1}(x)$, the exponential distribution.

The three distributions appearing in the conclusion of Theorem 3.2.1 are those which are the possible limiting distributions for normalized minima for independent and identically distributed random variables (see Chapter 2 in the book by J. Galambos (1978)). Theorem 2.1.2 and the characterization based on the property (P4*) are therefore integral parts of the asymptotic theory of extremes. Since we would like to avoid any overlap with any existing book, we refer the reader for those characterization theorems which follow from the asymptotic theory of extremes to the recent book by J. Galambos (1978).

Although the following theorem is related to a limit theorem, its proof does not require the general asymptotic theory of extremes. Both the result and the actual method of proof are due to B. Arnold (1971) and R.C. Gupta (1973).

Theorem 3.2.2. Assume that for an $n \geq 2$, $nX_{1:n}$ has the same distribution $F(x)$ as the population. If $F(x)$ is such that, as $x \to 0+$, $\lim F(x)/x = b > 0$ finite, then $F(x) = 1-e^{-bx}$, $x \geq 0$.

Proof: Let $n(k) = n^k$, $k \geq 1$, where $n \geq 2$ is fixed with the property of the theorem. If, in a set $X_1, X_2, \ldots, X_{n(k)}$, $k \geq 2$, of observations, we form blocks $(X_1, X_2, \ldots, X_{n(k-1)})$, $(X_{n(k-1)+1}, X_{n(k-1)+2}, \ldots, X_{2n(k-1)}), \ldots, (X_{(n-1)n(k-1)+1}, X_{(n-1)n(k-1)+2}, \ldots, X_{n(k)})$, and if the minima of these blocks are denoted by $X^{(1)}_{1:n(k-1)}, X^{(2)}_{1:n(k-1)}, \ldots, X^{(n)}_{1:n(k-1)}$, then evidently

$$X_{1:n(k)} = \min\{X^{(1)}_{1:n(k-1)}, X^{(2)}_{1:n(k-1)}, \ldots, X^{(n)}_{1:n(k-1)}\} . \tag{7}$$

Since the original X_j are independent and identically distributed, so are the $X^{(j)}_{1:n(k-1)}$. Hence, for k=2, each $X^{(j)}_{1:n(k-1)}$ is distributed as $X_{1:n}$ which by assumption has distribution $F(nx)$. Therefore, by the basic formula (2)

$$[1-F(x)]^n = 1-F(nx)$$

and thus by (7)

$$P(X_{1:n(2)} \geq x) = [1-F(nx)]^n = 1-F(n^2x) .$$

Formulas (2) and (7) and induction over k now yield

$$P(X_{1:n(k)} \geq x) = [1-F(x)]^{n(k)} = 1-F[n(k)x] .$$

Hence, on one hand

$$P(X_{1:n(k)} < \frac{x}{n(k)}) = 1-[1-F(\frac{x}{n(k)})]^{n(k)}$$

and on the other

$$P(X_{1:n(k)} < \frac{x}{n(k)}) = F(x) \ ,$$

where $n \geq 2$, and $n(k) = n^k$ with $k \geq 1$. We thus get

(8) $$F(x) = 1-[1-F(\frac{x}{n^k})]^{n^k} \quad \text{for all } k \geq 1 \ .$$

However, by assumption, as $k \to +\infty$, $F(xn^{-k}) = axn^{-k} + o(n^{-k})$. Consequently, the elementary relation

$$\lim_{s=+\infty} (1 + \frac{y}{s} + o(\frac{1}{s}))^s = e^y \ ,$$

and (8) imply that, for any $x > 0$,

$$F(x) = 1 - \lim_{k=+\infty}\left(1 - \frac{bx}{n^k} + o(\frac{1}{n^k})\right)^{n^k} = 1-e^{-bx} \ , \ x > 0 \ .$$

The proof is completed. □

The method of proof of Theorem 3.2.2 is a very useful one. It is called *the method of limit laws* in Galambos (1975a), because in this method we generated a sequence $n(k)$ from a single n, for which sequence a function had the same property for each k. Therefore, one can let k tend to infinity and this limit determines the fixed value expressed in the property valid for all k (in our case, formula (8)). We shall use this same method in the sequel in other proofs as well.

It is clear from the proof that (8) can be satisfied for functions other than the exponential if we do not assume that $F(x)/x \to a > 0$ as $x \to 0^+$. Namely, if we write $F(x) = 1-\exp[g(x)]$, where $g(x) < 0$ is nonincreasing then (8) requires that

(9) $$g(x) = n^k g(\frac{x}{n^k}) \quad \text{for all } x \text{ and all } k \geq 1 \ ,$$

where $n \geq 2$ is a fixed integer. J.S. Huang (1974) observes that such functions $g(x)$ can easily be constructed even under the additional assumption that $g(x)$ is differentiable. Out of the general class given by Huang we give here only one example. Let $g(x) = 0$ for all $x \leq 0$. If $x > 0$, define the integer (positive or negative) j such that $2^j < x \leq 2^{j+1}$. With this value of j we now choose

$$g(x) = -2^j\left(\frac{\log x - j \log 2}{\log 2} + 1\right) \ .$$

Here (9) evidently holds for $n = 2$; however, $F(x)$ is clearly not exponential.

Let us return to (P4*) once again. We can view this property as a converse of one part of Theorem 3.1.1, i.e., the condition that the distribution of $d_{0:n}$ is the same as that of the population is a characteristic property of the exponential distribution. There are evidently several other ways of formulating converse statements to parts of Theorem 3.1.1 and we can ask whether these are characteristic properties of the exponential distribution. Let us list some such possibilities:

(i) is it a characteristic property of the exponential distribution that $d_{r:n}$ is exponential?

(ii) is it a characteristic property of the exponential distribution that the distribution of $(n-r)d_{r:n}$ does not depend on r?

(iii) is it a characteristic property of the exponential distribution that $d_{r:n}$ and $d_{s:n}$ are independent for $r \neq s$?

We intentionally posed our questions vaguely without specifying requirements on r and n (whether just one or all values), which formulation allows us to refer to these questions after different specifications. We shall discuss questions (i) and (ii) here in this section and we devote a new section to question (iii).

In regard to question (i), we should emphasize that the case of $r = 0$ has been mentioned and the answer is almost trivial. Indeed, $d_{0:n} = X_{1:n}$, and we have remarked that, in view of (2), any preassigned distribution for $X_{1:n}$ uniquely determines the population distribution. Therefore, $r \geq 1$ only is of interest.

Now let $r = 1$ and $n = 2$. Then $d_{1:2} = X_{2:2} - X_{1:2} = |X_1-X_2|$. If the common distribution of X_1 and X_2 is exponential, then Theorem 3.1.1 implies that so is $d_{1:2}$. This fact, of course, can easily be computed by means of the convolution formula which yields that the density of X_1-X_2 equals $\frac{1}{2}e^{-|x|}$ from which we get the exponentiality of $d_{1:2} = |X_1-X_2|$. But if one starts with the distribution

$$F(x) = 1-e^{-x}[1 + 4b^{-2}(1-\cos bx)] \ , \ x \geq 0, \ b \geq 2\sqrt{2} \ ,$$

as the common distribution of X_1 and X_2, then the convolution formula leads to the same density $\frac{1}{2}e^{-|x|}$ for X_1-X_2 and thus $d_{1:2}$ again is exponential. Therefore, without further assumptions, the answer to question (i) is negative. The above example is due to H.-J. Rossberg (1972a), who also recognized the reason why distributions other than the exponential may have the property that $d_{r:n}$ is exponential. Namely, he realized that for all distributions F(x) with $F(0+) = 0$ for which $d_{r:n}$ is exponential, the Laplace transform

$$(10) \qquad u_r(s) = \int_0^{+\infty} e^{-sx} dF^r(x)$$

has zeros for some s such that $\operatorname{Re} s \geq 0$. Therefore, he posed the question whether under the assumption of $u_r(s) \neq 0$ for $\operatorname{Re} s \geq 0$, the exponentiality of $d_{r:n}$ implies that the population distribution is exponential. He proved the answer to this question in the affirmative by utilizing the following result.

Lemma 3.2.1. Let $R(t)$ be a real valued function of bounded variation on the nonnegative real line. Let $G(x)$ be a distribution function such that $G(0+) = 0$ and $G(x) > 0$ for all $x > 0$. Assume that

$$(11) \quad \int_0^{+\infty} G(t-x)dR(t) = 0 , \quad x \geq 0 ,$$

and that the Laplace transform

$$G^*(s) = \int_{-\infty}^{+\infty} e^{-sx}dG(x)$$

does not vanish for $\operatorname{Re} s \geq 0$. Then $R(t)$ is constant for $t > 0$.

Outline of proof: Let $R(+\infty)$ denote the limit of $R(t)$ at infinity. If we put $R_1(t) = R(t) - R(+\infty)$ then (11) means that the convolution $H(x)$ of $R_1(t)$ and $G(-t)$ is identically zero for $x \geq 0$. Therefore, the Laplace transform $H^*(s)$ of $H(x)$ is continuous and bounded for $\operatorname{Re} s \leq 0$. On the other hand, if $R_1^*(s)$ denotes the Laplace transform of $R_1(t)$, then $H^*(s) = -G^*(-s)R_1^*(s)$ for $\operatorname{Re} s = 0$ and $H^*(s)/G^*(-s)$ does not have singularities for $\operatorname{Re} s \leq 0$. From here one can easily conclude that $R_1^*(s)$ can be extended to $\operatorname{Re} s < 0$ by the formula $\{-H^*(s)/G^*(-s)\}$, which is bounded now on the whole complex plane. The basic theorem of Liouville on entire functions implies that $R_1^*(s)$ is constant. Hence, the uniqueness theorem of Laplace transforms implies that $R_1(t)$ also is constant. This completes the proof. □

We can now prove Rossberg's result.

Theorem 3.2.3. Let the population distribution $F(x)$ be such that $F(0+) = 0$ but $F(x) > 0$ for all $x > 0$. Assume that the Laplace transform $u_r(s)$ of (10) does not vanish for $\operatorname{Re} s \geq 0$. If, for some $r \geq 1$,

$$(12) \quad P(d_{r:n} \geq x) = e^{-(n-r)x} , \quad x \geq 0$$

then $F(x) = 1-e^{-x}$, $x \geq 0$.

Proof: By the method utilized in Theorem 3.1.3, we can prove

$$P(d_{r:n} \geq x) = -\binom{n}{r} \int_0^{+\infty} F^r(u-x)d[1-F(u)]^{n-r} .$$

Hence, by (12), for all $x \geq 0$,

$$-\binom{n}{r} \int_0^{+\infty} F^r(u-x)d[1-F(u)]^{n-r} = e^{-(n-r)x} ,$$

which can also be written as

$$\int_0^{+\infty} F^r(u-x)dR(u) = 0 , \quad x \geq 0 ,$$

where

$$R(u) = [1-F(u+y)]^{n-r} - e^{-(n-r)y}[1-F(u)]^{n-r} , \quad u > 0 ,$$

with some fixed $y > 0$. Hence, Lemma 3.2.1 yields that $R(u)$ is constant for $u > 0$. This constant value is evidently zero since $\lim R(u) = 0$ as $u \to +\infty$. That is,

$$\frac{1-F(u+y)}{1-F(u)} = e^{-y}, \quad \text{for all } y > 0 \text{ and } u > 0.$$

Letting $u \to 0$ we get $F(y) = 1-e^{-y}$, $y > 0$, which was to be proved. □

Let us turn to question (ii). There are only a few solutions to this problem under rather restrictive assumptions. Some of these solutions, however, are very valuable in engineering applications. Namely, several failure models can be approximated by a model in which the components are independent (but not identically distributed). For such models, however, the asymptotic extreme value distributions are of monotonic hazard rate (see J. Galambos (1978), Chapter 3, particularly Sections 3.9 through 3.12). Therefore, if one can characterize a distribution in the class of all distributions with monotonic hazard rate, then the general theory just mentioned combined with such a characterization theorem gives as good a practical solution as a mathematically more general characterization which assumes much less on the population distribution. Of course, for practical problems not related to the asymptotic theory of extremes, other kinds of assumptions are needed on the population. Hence, a solution to question (ii) would be of wider interest than the following result due to M. Ahsanullah (1976 and 1977a).

Theorem 3.2.4. Let $F(x)$ be absolutely continuous with density function $f(x) = F'(x)$. Assume that $F(0) = 0$, $F(x)$ is strictly increasing for all $x > 0$ and the hazard rate

$$H(x) = \frac{f(x)}{1-F(x)}$$

is monotonic for all $x \geq 0$. If for some $n \geq 2$, $nd_{0:n} = nX_{1:n}$ and $(n-1)d_{1:n}$ are identically distributed then $F(x) = 1-e^{-bx}$, $x \geq 0$, with some $b > 0$.

Notice that we are assuming only that two d's are identically distributed after normalization. It is not essential that we choose $nd_{0:n}$ and $(n-1)d_{1:n}$. In the proof below no modification is required if we assume that $(n-i)d_{i:n}$ and $(n-i+1)d_{i-1:n}$ are identically distributed or that $nd_{0:n}$ and $(n-i)d_{i:n}$ are identically distributed. These changes in the assumption would make the formulas in the proof algebraically but not conceptually more complicated, which is the sole reason for our choice of assumptions.

Proof: On account of (2),

$$(13) \quad P(nd_{0:n}<x) = 1-[1-F(\tfrac{x}{n})]^n .$$

When evaluating the distribution of $(n-1)d_{1:n}$, we condition it on $X_{1:n} = y$ and we use the continuous version of the total probability rule. We get

$$P[(n-1)d_{1:n}<x] = \int_0^{+\infty} P[(n-1)d_{1:n}<x \mid X_{1:n}=y] f_{1:n}(y)dy \; .$$

Now we substitute the expression (4) for $f_{1:n}(y)$ and we apply Theorem 3.1.3. We obtain

$$(14) \quad P[(n-1)d_{1:n}<x] = \int_0^{+\infty} \left[1 - \left(\frac{1-F(y+\frac{x}{n-1})}{1-F(y)} \right)^{n-1}\right] n[1-F(y)]^{n-1} f(y)dy$$

$$= 1 - n \int_0^{+\infty} \left[\frac{1-F(y+\frac{x}{n-1})}{1-F(y)} \right]^{n-1} [1-F(y)]^{n-1} f(y)dy \; .$$

If we write

$$[1-F(\tfrac{x}{n})]^n = n \int_0^{+\infty} [1-F(\tfrac{x}{n})]^n [1-F(y)]^{n-1} f(y)dy \; ,$$

(13) and (14) and the assumption that the distributions of $d_{0:n}$ and $d_{1:n}$ are the same yield

$$(15) \quad \int_0^{+\infty} K(x,y,n)[1-F(y)]^{n-1} f(y)dy = 0 \quad \text{for all } x \geq 0 \; ,$$

where

$$(16) \quad K(x,y,n) = [1-F(\tfrac{x}{n})]^n - \left[\frac{1-F(y+\frac{x}{n-1})}{1-F(y)} \right]^{n-1} .$$

Since the hazard rate $H(x) = -\frac{d}{dx}\log[1-F(x)]$ is monotonic, $\log[1-F(x)]$ is either concave or convex according as $H(x)$ is increasing or decreasing. For definiteness let $H(x)$ be increasing. Then the concavity of $\log[1-F(x)]$ implies

$$\log[1-F(y+\tfrac{x}{n})] \geq \frac{1}{n}\log[1-F(y)] + \frac{n-1}{n}\log[1-F(y+\tfrac{x}{n-1})] \; ,$$

which inequality is equivalent to

$$[1-F(y+\tfrac{x}{n})]^n \geq [1-F(y)][1-F(y+\tfrac{x}{n-1})]^{n-1} \; .$$

Therefore

$$(17) \quad K(x,y,n) \geq [1-F(\tfrac{x}{n})]^n - \left[\frac{1-F(y+\frac{x}{n})}{1-F(y)} \right]^n \geq 0 \; ,$$

where the last inequality again follows from the fact that $\log[1-F(x)]$ is concave. Now (17) implies that (15) can hold only if

$$K(x,y,n) = 0 \quad \text{for all } x,y \geq 0 \; .$$

That is,

$$(18) \quad 1-F(y+\tfrac{x}{n-1}) = [1-F(y)][1-F(\tfrac{x}{n})]^{n/(n-1)} \quad \text{for all } x,y \geq 0.$$

This is exactly the lack of memory property, because if we write $z = x/(n-1)$, (18) says that

$$\frac{1-F(y+z)}{1-F(y)}$$

does not depend on y. But in that case $y = 0$ gives that this fraction is necessarily $1-F(z)$. Since the lack of memory equation, when it is assumed to hold for all $y \geq 0$ and $z \geq 0$ has only two solutions (Theorem 1.3.1), namely, exponential or degenerate, our claim follows from the assumption that $F(x)$ is continuous.

The case when $H(x)$ is decreasing is similar to the above argument; we thus do not repeat it. This completes the proof. □

Our proof is somewhat different from the original proof given by Ahsanullah (1977a). With this new proof, we aimed to show that the assumptions of the theorem lead back to the basic properties described in Chapter 1. This is quite a surprising fact. We believe that the conclusion of the theorem can be shown in the above elementary way even without assuming the monotonicity of $H(x)$. We can prove several variants of the theorem, but we always needed some assumptions. We therefore left the formulation of the theorem in Ahsanullah's original form because of its practical value as explained earlier.

If one repeats the proof of Theorem 3.2.4 under the modified assumption that $nd_{0:n}$ and $(n-i)d_{i:n}$ are identically distributed then one immediately realizes that the exponential distribution has the following characteristic property: $d_{r:n}$ and $X_{1:n-r}$ are identically distributed. Since we conclude this fact from the proof of Theorem 3.2.4 (in its modified form), the additional assumptions (absolute continuity and monotonic hazard rate) are, of course, used in this characterization. These restrictions, however, are not necessary. H.-J. Rossberg (1972a) obtains the above characterization with the slight assumptions that $F(x)$ is not degenerate and is not a lattice distribution. M. Ahsanullah (1975) extends this type of characterization to a comparison of the distributions of $X_{s:n} - X_{r:n}$ and $X_{s-r:n-r}$ for two values of s, each of which is strictly larger than r. Rossberg's method is analytic on the line of the proof of Theorem 3.2.3. The basic idea of Ahsanullah is similar to his proof of Theorem 3.2.4, where the lack of monotonicity of $H(x)$ can be overcome by utilizing two values of s to derive the desired conclusion.

If we restate Rossberg's (1972a) theorem for $n = 2$, we get a result in which $d_{r:n}$ is compared with the population distribution $F(x)$. In fact, if $n = 2$, the statement that $d_{1:2}$ has the same distribution as $X_{1:1}$ simply means that with two independent observations X_1 and X_2, $d_{1:2} = |X_1-X_2|$ has the same distribution as the common distribution of X_1 and X_2. The fact that this property characterizes the exponential distribution (among continuous distributions) was obtained by P.S. Puri and H. Rubin (1970). Within a more restrictive family (which is, however, useful in engineering statistics), M. Ahsanullah (1977b) obtains that it is a characteristic property of the exponential distribution that $(n-r)d_{r:n}$ has the same distribution as the population. See also M. Ahsanullah and M. Rahman (1972) and J.S. Huang (1974a). For a characterization of the uniform distribution by distributional properties of $d_{r:n}$ of the type that it does not depend on r and F(x) is either super-additive or subadditive, see J.S. Huang, B.C. Arnold and M. Ghosh (1977).

3.3. Independence of functions of order statistics.

We present in this section a number of characterization theorems based on the independence of certain functions of order staitstics. In addition to the actual results, we also wish to draw attention to the methods of proof. The method based on Theorem 3.1.3 reduces characterizations by independence of functions of order statistics to some forms of the lack of memory property (Section 2.1). Another method utilizing characteristic functions has the potential of extending some results of the present section to stability theorems. We shall return to this possibility at the end of this section.

We start with the following result which was first proved by M. Fisz (1958) under somewhat different assumptions.

Theorem 3.3.1. Let X_1 and X_2 be independent random variables with common continuous distribution function F(x). Assume that F(0) = 0 and that F(x) is strictly increasing for all x > 0. Then $X_{2:2} - X_{1:2}$ and $X_{1:2}$ are independent if, and only if, $F(x) = 1-e^{-bx}$ with some b > 0.

Proof: The special case n = 2 of Theorem 3.1.1 implies that $X_{2:2} - X_{1:2}$ and $X_{1:2}$ are indeed independent for the exponential distribution $F(x) = 1-e^{-bx}$, $x \geq 0$, $b > 0$. Hence, only the converse statement needs proof.

If $X_{1:2}$ and $X_{2:2} - X_{1:2}$ are independent then

$$(19)\quad P(X_{2:2} - X_{1:2} < x | X_{1:2} = z) = P(X_{2:2} - X_{1:2} < x)$$

for almost all z > 0. Here, "almost all" can refer to Lebesgue measure because of the assumptions on F(x) that it is continuous and strictly increasing for all x > 0. On the other hand, by Theorem 3.1.3,

$$P(X_{2:2} - X_{1:2} < x | X_{1:2} = z) = P(X_{2:2} < x+z | X_{1:2} = z) = P(X^*_{1:1} < x+z) ,$$

where $X^*_{1:1}$ is the indicated order statistic from a population with parent distribution

$$F^*(x) = \begin{cases} \dfrac{F(x)-F(z)}{1-F(z)} & \text{if } x \geq z , \\ 0 & \text{otherwise} . \end{cases}$$

That is, if we denote the right hand side of (19) by $H(x)$, then

$$(20)\quad \frac{F(x+z) - F(z)}{1-F(z)} = H(x)$$

for all $x \geq 0$ and almost all $z > 0$. Letting $z \to 0$ implies that $F(x) = H(x)$. Thus, if we write

$$\frac{F(x+z) - F(z)}{1-F(z)} = 1 - \frac{1-F(x+z)}{1-F(z)} ,$$

(2) becomes

$$1 - F(x+z) = [1-F(x)][1-F(z)]$$

for all $x \geq 0$ and almost all $z > 0$. This is an extended form of the lack of memory property discussed in Section 2.1. In particular, we have from Theorem 2.1.1 that $F(x)$ is exponential as claimed. □

Although the assumptions of Theorem 3.3.1 are sufficient but not necessary for its conclusion, the following lemmas show that more general cases can easily be reduced to these assumptions.

Lemma 3.3.1. Let X_1 and X_2 be independent random variables with common nondegenerate distribution function $F(x)$. If $X_{1:2}$ and $X_{2:2} - X_{1:2}$ are independent then $F(x) < 1$ for all x.

Proof: Assume that there is a finite number A such that $F(A) = 1$. Then $X_1^* = X_1 - A$ and $X_2^* = X_2 - A$ are nonpositive with probability one and $X_{1:2}^*$ and $X_{2:2}^* - X_{1:2}^*$ are evidently independent, where a star in the superscript of the order statistics indicates that we are dealing with the order statistics of X_1^* and X_2^*. But for nonpositive values

$$X_{1:2}^* = X_{2:2}^* + X_{1:2}^* - X_{2:2}^* \leq X_{1:2}^* - X_{2:2}^*$$

with probability one, or equivalently, with probability one,

$$-X_{1:2}^* \geq X_{2:2}^* - X_{1:2}^* ,$$

which contradicts the independence of $X_{1:2}^*$ and $X_{2:2}^* - X_{1:2}^*$ and the assumption that $F(x)$ is nondegenerate. This completes the proof. □

Lemma 3.3.2. Under the assumptions of Lemma 3.3.1, there is a finite number B such that $F(B) = 0$.

Proof: Assume that the opposite is true. That is, for any finite number A, $F(A) > 0$ Then for any $A < 0$,

(21) $P(X_{1:2} < A,\ X_{2:2} - X_{1:2} \geq |A|) \geq 2P(X_1<A)P(X_2 \geq 0)$.

On the other hand, by independence,

(22) $P(X_{1:2} < A,\ X_{2:2} - X_{1:2} \geq |A|) = P(X_{1:2} < A)P(X_{2:2} - X_{1:2} \geq |A|)$

$$= [1 - (1-F(A))^2]P(X_{2:2} - X_{1:2} \geq |A|)$$

$$= F(A)[2 - F(A)]P(X_{2:2} - X_{1:2} \geq |A|) .$$

We thus get from (21) and (22) that

$$[1 - \tfrac{1}{2}F(A)]P(X_{2:2} - X_{1:2} \geq |A|) \geq P(X_2 \geq 0) .$$

Letting $A \to -\infty$ we obtain $P(X_2 \geq 0) = 0$, which contradicts Lemma 3.3.1. The lemma is established. □

Lemma 3.3.3. Let X_1 and X_2 be independent and identically distributed geometric variables, i.e.

$$P(X_1=k) = P(X_2=k) = p^k(1-p) ,\ 0 < p < 1,\ k = 0,1,2,\ldots .$$

Then $X_{1:2}$ and $X_{2:2} - X_{1:2}$ are independent.

Proof: Since, for any integers $n \geq 0$ and $t \geq 0$, satisfying $n+t > 0$,

$$P(X_{2:2} - X_{1:2} = n,\ X_{1:2} = t) = 2P(X_1 = t,\ X_2 = n+t)$$

$$= 2p^t(1-p)p^{n+t}(1-p) = 2p^{n+2t}(1-p)^2 ,$$

and moreover,

$$P(X_{2:2} - X_{1:2} = 0,\ X_{1:2} = 0) = P(X_1 = X_2 = 0) = (1-p)^2 ,$$

we immediately get that $X_{1:2}$ and $X_{2:2} - X_{1:2}$ are independent. □

Lemma 3.3.4. Let the conditions of Lemma 3.3.1 be fulfilled; assume also that $F(x)$ has the following property: there are two numbers $a < b$ such that $0 < F(a) = F(b)$. Then $F(x)$ is not continuous.

Proof: We again assume that the opposite is true, i.e., $F(x)$ is continuous but not strictly increasing on a semiline. In other words, there are $a < b$ such that $F(a) > 0$ and $F(b) = F(a)$. We can, of course, choose a so that $F(z) < F(a)$ for all $z < a$. Furthermore, because $F(\cdot)$ is continuous, we can choose two values $z_1 < z_2$

such that $0 < F(z_1) < F(z_2) < F(a)$ and $a - z_1 < \varepsilon$, where $\varepsilon > 0$ is arbitrary. We shall impose one more restriction on the choice of z_1 and z_2 later.

Now if we start with (19), we again get (20), but the condition that "almost all $z > 0$" refers here to the measure $Q(D)$ on the Borel sets D of the real line for which $Q(D) = F(d) - F(c)$ for intervals $D = (c,d)$. Hence, (20) may not hold for $a < z < b$, where $a < b$ are as defined above. But, if $D = (a-\varepsilon,a)$, then by the choice of a, $Q(D) = F(a) - F(a-\varepsilon) > 0$, and thus we can choose two points $z_1 < z_2$ in this interval D for which (20) holds with all $x \geq 0$. Therefore, if we substitute first $z = z_2$ and $x = a-z_2$ and then $z = z_2$ and $x = b-z_2$ into (20) and subtract the two equations, the left hand sides cancel out and thus (20) implies

$$0 = H(a-z_2) - H(b-z_2) .$$

Since $H(x)$ denotes the distribution function on the right hand side of (19), this means that

$$P(a-z_2 \leq X_{2:2} - X_{1:2} < b-z_2) = 0 .$$

This last equation, however, contradicts the fact that $P(z_1 \leq X_1 < z_2) = F(z_2) - F(z_1) > 0$ and $P(z_2 \leq X_2 < a) = F(a) - F(z_2) > 0$ and that $0 < a-z_1 < \varepsilon$, where $\varepsilon > 0$ is arbitrary. Namely, since $F(x)$ is continuous, it has positive probability that X_1 is between z_1 and z_2 and X_2 is arbitrarily close to a. For such values of X_1 and X_2, however, $X_{2:2} - X_{1:2} = X_2 - X_1$ is larger than $a-z_2$ and is smaller than $\varepsilon > 0$, which was arbitrary (hence, smaller than $b-z_2$). We have thus obtained that it is impossible to have $0 < F(a) = F(b)$ for some $a < b$ for continuous distribution functions $F(x)$ satisfying the conditions of the Lemma. Hence, the proof is complete. □

A careful analysis of the preceding proof shows that the following more general result is valid.

Lemma 3.3.5. If the conditions of Lemma 3.3.4 are satisfied then $F(x)$ is discrete.

Since we do not need any essential modification of the preceding proof, we omit details.

We now summarize Lemmas 3.3.1 - 3.3.5 and Theorem 3.3.1 as

Theorem 3.3.2. Let X_1 and X_2 be independent random variables with common distribution function $F(x)$. If $X_{1:2}$ and $X_{2:2} - X_{1:2}$ are independent, then $F(x)$ is either discrete or $F(x) = 1 - \exp[-b(x-B)]$, $x \geq B$, where $b > 0$ and B are finite constants.

We have seen in Lemma 3.3.3, that there are discrete distributions for which $X_{1:2}$ and $X_{2:2} - X_{1:2}$ are independent. One additional class of discrete distributions is the degenerate one for which the above independence property still holds. It is, however, true that there are no other discrete distributions with this independence property. Since we do not analyze in detail the discrete distributions, we do not prove this claim but we add that the support of the geometric distribution does not necessarily have to be the set of the nonnegative integers as it was specified in Lemma 3.3.3.

Several extensions of Theorem 3.3.2 are known in the literature. However, before formulating some of these extensions, we give some references. Although the proofs presented here are new, different versions of Theorem 3.3.2, and thus of the lemmas as well, can be found in several publications. As usual, we do not make a distinction between two results if they differ only in a monotonic transformation (see for example formulas (17) and (18) in Section 1.6). As mentioned earlier, the first result along this line is due to M. Fisz (1958), who assumes absolute continuity and that $F(x)$ is not positive on the whole real line. H.-J. Rossberg (1960) relaxes the condition to continuity, for which case Z. Govindarajulu (1966) gives a new proof. Govindarajulu also investigates several other characterizations based on the independence of some linear functions of order statistics. One of them is an extension of a result by E.A. Tanis (1964), who proved that, among absolutely continuous distributions, only the exponential one possesses the property that

$$S_n = \sum_{j=1}^{n} X_j - nX_{1:n} = \sum_{j=1}^{n} (X_{j:n} - X_{1:n})$$

and $X_{1:n}$ are independent. Notice that S_n can also be written as

$$S_n = \sum_{j=1}^{n} c_j X_{j:n}, \quad \sum_{j=1}^{n} c_j = 0, \; c_1 \neq 0,$$

where the coefficients c_j are constants, (in fact, they are specified above). Guided by this observation, Rossberg (1972b) established the following very general extension of Theorem 3.3.1, proof of which will be discussed later.

<u>Theorem 3.3.3</u>. Let $X_1, X_2, \ldots, X_n$ be independent random variables with common continuous distribution function $F(x)$. Let $c_k, c_{k+1}, \ldots, c_n$ be constants satisfying $c_k \neq 0$, and $c_k + c_{k+1} + \ldots + c_n = 0$. Then $X_{k:n}$ and $S_{k,n} = c_k X_{k:n} + \ldots + c_n X_{n:n}$ are independent if, and only if, $F(x) = 1 - \exp[-b(x-B)]$, $x \geq B$, where $b > 0$ and B are finite constants.

Some variants of this theorem, under the assumption of absolute continuity, are also stated by A.P. Basu (1965).

Theorem 3.3.2 follows from results by T.S. Ferguson (1967) and J.H.B. Kemperman (1971). They actually determine all distributions F and G under which $\min(X_1, X_2)$ and $|X_1 - X_2|$ are independent, where X_1 and X_2 are independent random variables with distribution functions F and G, respectively. Earlier, under the stronger assumption of $\min(X_1, X_2)$ and $X_1 - X_2$ being independent, Ferguson (1964) and (1965) and G.B. Crawford (1966) found that necessarily F and G are either both exponential or both geometric (see Chapter 6 for further details of the results by Ferguson, Crawford and Kemperman).

J. Galambos (1972 and 1975c) proved two additional extensions of Fisz's theorem. In the first one, the independence of $X_{1:2}$ and $g(X_{1:2}, X_{2:2})$ is assumed, where $g(x,y) \geq 0$ is a Borel measurable function (see Theorem 3.3.4). In the second, the independence of $X_{1:2}$ and $X_{2:2} - X_{1:2}$, truncated at a given value, is assumed, and the conclusion is again that the parent distribution is exponential. This latter characterization is of significant practical value for the following reason. In many practical situations when it is expensive to wait for random occurrences we can save by comparing $X_{1:2}$ and the time elapsed until the second "occurrence" $X_{2:2}$ or until a given time, whichever occurs first, rather than measuring the actual value of $X_{2:2}$. This is a continuous analog of R.C. Srivastava's (1974) result for the geometric distribution. Namely, Srivastava obtained that if the parent distribution $F(x)$ is discrete and if $X_{1:2}$ is independent of the event $\{X_{2:2}-X_{1:2} = 0\}$ (which can be considered as a truncated version of $X_{2:2} - X_{1:2}$), then $F(x)$ is a geometric distribution function in the extended sense that the set of jumps of $F(x)$ is not specified.

Rossberg uses characteristic functions in his proof of Theorem 3.3.3. The method of characteristic functions is also the basic tool of Galambos (1972) and, for another class of problems, of Rossberg (1968). We shall illustrate this method by means of one of the quoted results of Galambos (1972). The advantage of the method of characteristic functions is that it is a promising avenue for extending a characterization theorem to a stability theorem. Namely, if instead of independence, we assume only almost independence in a well defined sense, then the deviation of the joint characteristic function from the corresponding product (which would have been obtained under independence) can be estimated. On the other hand, there are standard methods in probability theory to estimate the difference of two distribution functions when an estimate is available for the difference of their characteristic functions. Such a method is well exemplified by some proofs in the monograph by E. Lukacs and R.G. Laha (1964) (see pp. 150-170). It is an important and challenging area of research to extend the results of the present section to stability theorems.

<u>Theorem 3.3.4</u>. Let $g(x,y) \geq 0$ be a Borel measurable function. Let X_1 and X_2 be independent random variables with common absolutely continuous distribution function $F(x)$. Assume that $U = g(X_{1:2}, X_{2:2})$ has finite expectation. If $X_{1:2}$ and U are independent, then there is a constant c such that

$$(23) \quad \int_z^{+\infty} g(z,x)f(x)dx = c[1-F(x)]$$

for all z, where $f(x) = F'(x)$.

Proof: We use characteristic functions. For the basic properties of characteristic functions that we apply here, see for example, E. Lukacs and R.G. Laha (1964, pp. 23-24). Let the characteristic functions of the variables $X_{1:2}$, U and the vector $(X_{1:2},U)$ be denoted by $\phi_1(t)$, $\phi_2(u)$ and $\phi(t,u)$, respectively. Then the independence of $X_{1:2}$ and U is equivalent to the validity of the equation

$$(24) \quad \phi(t,u) = \phi_1(t)\phi_2(u)$$

for all real numbers t and u. Now the joint density of $(X_{1:2},X_{2:2})$ is $2f(x)f(y)$ for $x < y$, and zero otherwise. Hence

$$\phi(t,u) = 2\int_{-\infty}^{+\infty}\int_{x}^{+\infty} \exp\{itx + iug(x,y)\}f(x)f(y)dxdy ,$$

which, after the substitution $x+v = y$, becomes

$$\phi(t,u) = 2\int_{-\infty}^{+\infty}\int_{0}^{+\infty} \exp\{itx + iug(x,x+v)\}f(x)f(x+v)dvdx .$$

A similar calculation yields

$$\phi_1(t) = 2\int_{-\infty}^{+\infty}\int_{0}^{+\infty} e^{itx}f(x)f(x+v)dvdx$$

and

$$\phi_2(u) = 2\int_{-\infty}^{+\infty}\int_{0}^{+\infty} \exp\{iug(x,x+v)\}f(x)f(x+v)dvdx .$$

If we write these explicit expressions in (24), then differentiate with respect to u and set $u = 0$, we get

$$\int_{-\infty}^{+\infty}\int_{0}^{+\infty} e^{itx}g(x,x+v)f(x)f(x+v)dvdx = c\int_{-\infty}^{+\infty}\int_{0}^{+\infty} e^{itx}f(x)f(x+v)dvdx .$$

Since $g(x,x+v) \geq 0$ by assumption, the inversion formula for characteristic functions implies that

$$\int_{0}^{+\infty} g(x,x+v)f(x)f(x+v)dv = c\int_{0}^{+\infty} f(x+v)dv .$$

This equation is equivalent to (23) (substitute $z = x+v$), and thus the proof is complete. □

Notice that if $g(x,y) = y-x$ for $y \geq x$ and zero otherwise, then Theorem 3.3.4 reduces to the Fisz theorem. Indeed, in this case (23) takes the form

$$\int^{+\infty} (x-z)f(x)dx = c[1-F(z)]$$

or equivalently

$$\int_z^{+\infty} xf(x)dx - z[1-F(z)] = c[1-F(z)] .$$

Differentiating with respect to z, we have

$$1 - F(z) = cf(z) ,$$

which evidently leads to the exponential distribution.

Other applications are immediate. The reader is invited to write down some special cases by choosing other explicit forms for $g(x,y)$. In particular, it will be interesting to work out special cases for which it is not straightforward that the resulting distribution is a simple transform of the exponential distribution.

Some of the results of the next section, namely, those characterizations which are based on constant regression, are closely related to the present section. They are not necessarily generalizations, however, because the assumptions will differ, mainly by imposing finiteness of expectations, which was not a standard assumption in this section.

3.4. Characterizations through moment assumptions.

Let $X_1, X_2, \ldots, X_n$ be independent random variables with common distribution function $F(x)$. We introduce the notation

$$(25) \quad E_{r:n} = E(X_{r:n}) ,$$

which, when used, is always assumed to be finite. In view of (3),

$$(26) \quad E_{r:n} = \int_{-\infty}^{+\infty} xdF_{r:n}(x) = r\binom{n}{r} \int_0^1 F^{-1}(u)u^{r-1}(1-u)^{n-r}du ,$$

whenever the right hand side is finite, where

$$F^{-1}(u) = \inf\{x : F(x) \geq u\} .$$

We thus have from (26) and the elementary relation

$$(n-r)\binom{n}{r}(1-u) + (r+1)\binom{n}{r+1}u = n\binom{n-1}{r} ,$$

that, for any integers $0 < r < n$, $n \geq 2$,

$$(27) \quad (n-r)E_{r:n} + rE_{r+1:n} = nE_{r:n-1} .$$

Hence, if $r = r(n)$ is an arbitrary function of n such that $r(n)$ is an integer with $1 \leq r(n) \leq n$, then the sequence $E_{r(n):n}$, $n \geq 1$, where $E_{1:1} = E(X_1)$, uniquely determines all values $E_{r:n}$, $1 \le r \le n$, $n \ge 1$, on account of (27). We thus obtained the following basic result, which was first pointed out independently by J.B. Kadane (1974) and by Galambos (1975a).

Lemma 3.4.1. If the integers $r = r(n)$ satisfy $1 \le r(n) \le n$ for all $n \ge 1$, then the following two statements are equivalent:

(i) property A holds for the sequence $E_{r(n):n}$, $n \ge 1$,

(ii) property A holds for all values $E_{r:n}$, $1 \le r \le n$, $n \ge 1$,

where $E_{1:1} = E(X_1)$.

As a special case, we get from Lemma 3.4.1, that statements such as "$E_{1:n}$, $n \ge 1$, uniquely determines the population distribution" or "the triangular array $E_{r:n}$, $1 \le r \le n$, $n \ge 1$, of numbers uniquely determines the population distribution" are equivalent. Therefore, the following result, due to W. Hoeffding (1953), is as general as subsequent "generalizations" (L.K. Chan (1967), A.G. Konheim (1971), M. Pollack (1973), R.C. Gupta (1974) and M.M. Ali (1976)).

Theorem 3.4.1. The triangular array $E_{r:n}$, $1 \le r \le n$, $n \ge 1$, of numbers uniquely determines the population distribution $F(x)$.

Proof: Let $Y_1, Y_2, \ldots, Y_n$ be i.i.d. random variables with distribution function $T(x)$. Assume that, for all $1 \le r \le n$, $n \ge 1$,

$$E(Y_{r:n}) = E_{r:n} ,$$

where $E_{r:n}$ is as defined at (25) for the basic sequence $X_1, X_2, \ldots, X_n$ (recall that the notation $E_{r:n}$ automatically assumes that this quantity is finite). Utilizing now (26), we get

$$\text{(28a)} \quad \int_0^1 F^{-1}(u)u^{r-1}(1-r)^{n-r}du = \int_0^1 T^{-1}(u)u^{r-1}(1-u)^{n-r}du$$

for all $1 \le r \le n$, $n \ge 1$. Let $r \ge 1$ be fixed and consider the last equation for all $n \ge r$. Substituting $y = 1-u$ and putting $f_r(y) = F^{-1}(1-y)(1-y)^{r-1}$ and $t_r(y) = T^{-1}(1-y)(1-y)^{r-1}$ results in

$$\text{(28)} \quad \int_0^1 f_r(y)y^k dy = \int_0^1 t_r(y)y^k dy , \quad k = n-r \ge 0 .$$

Since both $f_r(y)$ and $t_r(y)$ are nonnegative and their integrals are equal (set $k = 0$ in the above equation) there exists a constant $c > 0$ such that both $cf_r(y)$ and $ct_r(y)$ are densities over the finite interval $(0,1)$. Hence (28) says that all moments of two absolutely continuous distributions are equal, which distributions are supported by the finite interval $(0,1)$. The classical moment problem (see, e.g., J. Galambos (1978), Appendix II) thus implies that $f_r(y) = t_r(y)$ which in turn yields $F(x) = T(x)$ as claimed. □

This theorem admits many interesting corollaries. For example

Corollary 3.4.1. If $E_{1:n} = 1/n$ for all $n \geq 1$, then $F(x) = 1-e^{-x}$, $x \geq 0$.

Proof: If $F(x) = 1-e^{-x}$, $x \geq 0$, then $E_{1:n} = 1/n$ for all $n \geq 1$. Thus, by Lemma 3.4.1, $E_{r:n}$ is the same for all $1 \leq r \leq n$, $n \geq 1$, as for a population distribution $F(x) = 1-e^{-x}$. Theorem 3.4.1 now implies our statement. □

Corollary 3.4.2. If $E_{1:n} = 1/(n+1)$, $n \geq 1$, then $F(x) = x$ for $0 \leq x \leq 1$.

Proof: The argument is similar to the preceding one. Since, with $F(x) = x$, $0 \leq x \leq 1$, $E_{1:n} = 1/(n+1)$, Lemma 3.4.1 and Theorem 3.4.1 imply that no other population distribution can have this property. □

It is evident from Corollaries 3.4.1 and 3.4.2 that the limits of $nE_{1:n}$, as $n \to +\infty$, cannot characterize the population distribution. Galambos (1975a) gives an heuristic argument to show that asymptotic values $E_{k:n} \sim h(k,n)$, $1 \leq k \leq n$, $n \to +\infty$, may characterize population distributions within some families. It would be interesting to make that argument rigorous.

We state another consequence of Corollary 3.4.2. Let $Y_1, Y_2, \ldots, Y_{n+1}$ be i.i.d. nonnegative random variables. Define $S_j = Y_1+Y_2+\ldots+Y_j$ and consider the sequence $R_{j,n} = S_j/S_{n+1}$. We then have

Corollary 3.4.3. If the random variables $R_{j,n}$, $1 \leq j \leq n$, are distributed as the order statistics $X_{j:n}$, $1 \leq j \leq n$, of a sample from a population with parent distribution $F(x)$, then $F(x) = x$, $0 \leq x \leq 1$.

Proof: Since $Y_j \geq 0$, the random variables $R_{j,n}$ are bounded. Hence, $E(R_{j,n})$ is finite. Furthermore, $E(Y_j/S_{n+1})$ does not depend on j and $\sum_{j=1}^{n+1} E(Y_j/S_{n+1}) = 1$. Consequently,

$$E(R_{1,n}) = E(Y_1/S_{n+1}) = \frac{1}{n+1} \sum_{j=1}^{n+1} E(Y_j/S_{n+1}) = \frac{1}{n+1} .$$

Now, if $R_{j,n}$, $1 \leq j \leq n$, are the order statistics of a sample from a population with parent distribution $F(x)$, then Corollary 3.4.2 is applicable, whose conclusion is exactly our claim. □

The statistics $R_{j,n}$, $1 \leq j \leq n$, of the sample $Y_1, Y_2, \ldots, Y_{n+1}$ are widely used in goodness of fit tests. Namely, if the common distribution of the Y's is $F(x) = 1-e^{-bx}$, then the joint distribution of the $R_{j,n}$ is indeed the same as that of the order statistics of a sample from a uniform population on $(0,1)$. The advantage of the transformation $R_{j,n}$ is that it does not depend on the unknown parameter b of $F(x)$. We shall return to this subject matter, and the corresponding characterization theorems, in Chapter 6.

The following interesting fact was observed by J.S. Huang (1974b) in connection with Corollary 3.4.1. If we write $E_{1:n} = 1/n$ in the form of $E(nX_{1:n}) = E(X_1)$ (which we assumed to be one in Corollary 3.4.1), then the condition of Corollary 3.4.1 becomes strongly related to the property (P4*) discussed in Section 1.5. The seemingly additional assumption here, compared with (P4*), is that $E(X_1)$ is assumed to be finite. Now, Huang observed that if (P4*) holds then $E(X_1)$ is finite and thus Corollary 3.4.1 is a much stronger statement than (P4*).

Let us return to Theorem 3.4.1. The question arises of how large should a subset of the numbers $E_{r:n}$, $1 \le r \le n$, $n \ge 1$, be in order to characterize the population distribution? In view of Lemma 3.4.1, a reduction in the set of values of r alone does not lead to stronger results, hence a meaningful extension of Theorem 3.4.1 can only be obtained by considering subsets of all the consecutive integers $n \ge 1$. Such an extension can indeed be achieved if we reexamine the essential step in the proof of Theorem 3.4.1. Namely, we obtained at (28) that Theorem 3.4.1 is essentially equivalent to the classical moment problem of bounded random variables. This, in turn, can be considered from the point of view of orthogonality or completeness of a set of functions by writing (28a) as

$$(29)\quad \int_0^1 [F^{-1}(u) - T^{-1}(u)]u^{r-1}(1-u)^{n-r}du = 0$$

for all $1 \le r \le n$, $n \ge 1$. Since it is known in functional analysis, that only the identically zero function can be orthogonal to the set of polynomials $u^{r-1}(1-u)^{n-r}$, $1 \le r \le n$, $n \ge 1$, it follows that $F^{-1}(u) = T^{-1}(u)$ and thus $F(u) = T(u)$ (we can easily get rid of the restriction that statements are valid only almost surely with respect to Lebesgue measure). This theorem is known as the completeness of the above polynomials over the interval (0,1). The question is therefore which subset of the above polynomials is complete over the interval (0,1)? We quote, without proof, one solution to this question as proposed by J.S. Huang (1975).

Theorem 3.4.2. Let the sequence $r(n)$ of integers be such that $1 \le r(n) \le n$ and, with some fixed $m \ge 1$, $r(m) \le r(n) \le r(m)+n-m$ for all $n \ge m$. Then if $E(X_{r(m):m})$ is finite, the sequence $E_{r(n):n}$, $n \ge m$, uniquely determines the population distribution $F(x)$.

As an example, consider the Pareto distribution $F(x) = 1-1/x$, $x \ge 1$. If $X_1, X_2, \ldots, X_n$ are i.i.d. with this distribution $F(x)$, then Theorem 3.4.1 is not applicable, because $E(X_1) = E_{1:1} = +\infty$. But, since $E_{1:2}$ is finite, Theorem 3.4.2 implies that $E_{r(n):n}$, $n \ge 2$, characterizes the Pareto distribution $F(x) = 1-1/x$, $x \ge 1$, whenever $1 \le r(n) \le n-1$ for $n \ge 2$ (we chose $m = 2$ and $r(m) = 1$).

In this same paper, Huang (1975) points out that a certain population distribution may not be characterized by the sequence $E_{r(n):n}$ simply because its members are not finite for any n, while through an appropriate monotonic transformation this same distribution can be characterized by a moment sequence (recall that any continuous distribution can be transformed into exponential). Such an example is $F(x) = 1-1/\log x$, $x \ge e$. Indeed, by virtue of (2),

$$P(X_{1:n}<x) = 1 - (\log x)^{-n} ,$$

and thus, for any $r \ge 1$,

$$E_{r:n} \ge E_{1:n} = \int_{e}^{+\infty} [1-P(X_{1:n}<x)]dx = \int_{e}^{+\infty} \frac{dx}{(\log x)^n} = +\infty .$$

However, with the transformed sequence $\log\log X_{r:n} = X^*_{r:n}$, Corollary 3.4.1 is applicable and we get that if, for all $n \ge 1$, $E(\log\log X_{1:n}) = 1/n$ then the population distribution is $F(x) = 1-1/\log x$, $x \ge e$.

Although we stated our theorems in terms of the first moments of order statistics, they are directly extendable to higher moments as well. For example, if the basic random variables X_j are nonnegative then the monotonic transformation $X_j^t = X_j^*$ transforms the order statistics $X_{r:n}$ to $X^t_{r:n}$, $t > 0$, and thus Theorem 3.4.1, applied to X_j^*, implies that the numbers $E(X^t_{r:n})$, $1 \le r \le n$, $n \ge 1$, uniquely determine the population distribution $F(x)$, whenever $E(X_1^t)$ is finite. If $X_j \ge 0$ is not assumed then the argument still works if t is an odd integer.

We have to be careful with subsequences $E_{r(n_k):n_k}$, however. Huang (1975) gives the example that if $F(x)$ is symmetric about the origin then $E(X_1) = 0$ and in fact $E_{r(n):n} = 0$ for all odd n and $r(n) = \frac{1}{2}(n+1)$. Therefore, this particular subsequence evidently does not characterize any distribution. On the other hand, if we take extreme order statistics rather than quantiles, then $E_{r(n_k):n_k}$ does characterize the population distribution for a large class of subsequences. Namely, the following result holds (see Huang (1975)).

Theorem 3.4.3. Assume that there is a pair (t,m) of integers such that $1 \le t \le m$ and $E_{t:m}$ is finite. Then $E_{r(n_k):n_k}$, $k \ge 1$, uniquely determines the population distribution whenever $1 \le r(n_k) \le t$ and $n_k \ge m$ for all $k \ge 1$, and if

$$\sum_{k=1}^{+\infty} \frac{1}{n_k} = +\infty .$$

It is, of course, evident that we could use upper extremes as well, i.e. $0 \le n_k - r(n_k) \le t$.

As the case of Theorem 3.4.2, this theorem also reduces to a completeness theorem for the polynomials appearing in (29). Such a completeness theorem can be found for example in the book by R.P. Boas (1954, p. 235), which we do not reproduce here.

For further results on subsets of the numbers $E_{r:n}$, $1 \le r \le n$, $n \ge 1$, which characterize the population distribution, see B.C. Arnold and G. Meeden (1975). Although Arnold and Meeden consider arbitrary moments (not just first ones), most of their results can be given in an equivalent form in terms of first moments, as was remarked in the second paragraph preceding Theorem 3.4.3.

The investigation by Arnold and Meeden is related to the following area of research developed by J.B. Kadane (1971 and 1974) and by C. Mallows (1973). Let $A = \{a_{r,n}: 1 \le r \le n, 1 \le n \le N\}$ be a finite triangular array of numbers. What conditions on A can guarantee that there exists a distribution function F(x) concentrated on the nonnegative real line such that, if $X_1, X_2, \ldots, X_n$ are i.i.d. random variables with common distribution function F(x), then $E_{r:n} = a_{r,n}$ for all entries of A? In their solutions, both Kadane and Mallows reduce this problem to a moment problem, combined with a recursive formula similar to (27). The reduction of this problem to a moment problem can, of course, go on the line of our arriving at the equation (28), but the emphasis is different here. Not so much the uniqueness of F(x) is emphasized, but rather the existence of at least one F(x) with the above property when A is given. Kadane (1974) observes that a recursive formula similar to (27) makes it possible that only the last entries $a_{n,n}$ in each column are needed to decide the existence of a distribution function F(x) with the previously stated properties. His result which also contains the earlier result of Mallows, is as follows. If the elements of A satisfy the quoted recursive relation and if the numbers $m_k = d_k$, $k = 0,1,\ldots,N-2$, where $d_n = a_{n,n} - a_{n-1,n-1}$, $2 \le n \le N$, are the k-th moments of a positive measure on (0,1), then there is a distribution function such that $E_{r:n} = a_{r,n}$ for each entry of A. This, of course, becomes a characterization theorem whenever the moments m_k characterize a distribution. It is very rare with finite N. However, if N is infinite then Theorem 3.4.1 implies that, for a given A, either there is a unique distribution F(x) with the above properties or there is none.

Let us now return to problems of characterization by the set $E_{r:n}$ of first moments.

Let us introduce the differences $\Delta_{r:n} = E_{r:n} - E_{r-1:n}$, $1 \le r \le n$, $n \ge 1$, where $E_{0:n} = 0$. It is, of course, equivalent to specify $\Delta_{r:n}$ or $E_{r:n}$ for all $1 \le r \le n$ and $n \ge 1$. However, if only some values of r are considered for each n, then it is a new problem whether the sequence $\Delta_{r(n):n}$, $n \ge 1$, characterizes the population distribution. In this connection, the most general known result is due to Z. Govindarajulu *et al.* (1975), which we state below.

Theorem 3.4.4. Let $X_1, X_2, \ldots, X_n$ be independent random variables with common distribution function F(x). Assume that there are two integers $1 \le k < m$ such that both $E_{k:m}$ and $E_{k+1:m}$ are finite. Then the sequence $\Delta_{k+1:n}$, $n \ge m$, uniquely determines the family $\{F(x+c)\}$ of distribution functions, where c is an arbitrary real number.

Remark: Since upper extremes can be transformed into lower extremes by turning to the sequence $\{-X_j\}$, it follows that the sequence $\Delta_{n-k:n}$, $n \geq m$, also determines the family $F(x+c)$, given that both $E_{m-k:m}$ and $E_{m-k-1:m}$ are finite for some k and m.

Proof: Since the proof is exactly the same as that of Theorem 3.4.1, we therefore give a short outline only. If we start with (26) again, then with the notations of the proof of Theorem 3.4.1, we get

$$(30) \quad \int_0^1 [F^{-1}(u) - T^{-1}(u)](nu-k)u^{k-1}(1-u)^{n-k-1}du = 0 , \quad n \geq m .$$

Because now the multiplier of $(1-u)^{n-k-1}$ depends on the exponent n, one additional trick is needed to reduce (30) to a completeness theorem as compared with the case of (29). The major idea of Govindarajulu *et al.* (1975) is to recognize that with the beta function

$$B_{a,b}(u) = \frac{\Gamma(a+b)}{\Gamma(a)\Gamma(b)} u^{a-1}(1-u)^{b-1}$$

and with the constant

$$c = \int_0^1 [F^{-1}(u) - T^{-1}(u)]B_{k,m-k}(u)du ,$$

(30) is equivalent to

$$\int_0^1 [F^{-1}(u) - T^{-1}(u) - c]u^{k-1}(1-u)^{j-1}du = 0 \quad \text{for all } j \geq m-k ,$$

which can also be written as

$$(31) \quad \int_0^1 f_k(u)u^s du = 0 , \quad s = 0,1,2,\dots ,$$

where

$$f_k(u) = [F^{-1}(u) - T^{-1}(u) - c]u^{k-1}(1-u)^{m-k-1} .$$

Now, just as in the proof of Theorem 3.4.1, (31) implies that $f_k(u) = 0$ for almost all u. Since F and T are distribution functions, we obtained that $T(x) = F(x+c)$ as claimed. □

For characterizations by sequences of variances and covariances of order statistics, see Z. Govindarajulu (1966 and 1975).

The results presented so far in this section reduce distributional properties to expectations. On the other hand, if we want to simplify characterizations based on independence, then conditional expectations are the natural tools to be used. We devote the rest of this section to such characterizations, i.e., when our assumptions are in terms of conditional expectations (regression) of functions of order statistics. Such results are not necessarily extensions of those in Section 3.3, since the finiteness of conditional expectations presumes that the corresponding expectations (without condition) are also finite, which was not assumed in Section 3.3. However, the actual results are easier to apply in many cases.

Again, let $X_1, X_2, \ldots, X_n$ be independent random variables with common distribution function $F(x)$. Motivated by the theorem of Fisz (Theorem 3.3.1), we first prove the following result.

Theorem 3.4.5. If $E(X_1)$ is finite and if $F(x)$ is continuous, then $E(X_{2:2}-X_{1:2}|X_{1:2} = y)$ is constant almost surely with respect to $F(x)$ if, and only if, $F(x)$ is exponential.

Proof: Since, for any y such that $0 < F(y) < 1$,

$$E(X_{2:2}-X_{1:2}|X_{1:2} = y) = E(X_{2:2}|X_{1:2} = y) - y ,$$

Theorem 3.1.3 implies that, for almost all y with respect to $F(y)$, for which $0 < F(y) < 1$,

$$(32) \quad E(X_{2:2}-X_{1:2}|X_{1:2} = y) = \int_y^{+\infty} x dF^*(x) - y ,$$

where

$$(33) \quad F^*(x) = \frac{F(x)-F(y)}{1-F(y)} \quad \text{for } x \geq y .$$

If the left hand side of (32) is constant almost surely then (32) is equivalent to

$$\int_y^{+\infty} x dF^*(x) = c + y \quad \text{a.s. (with respect to F)}$$

or to

$$(34) \quad \int_y^{+\infty} x dF(x) = (c+y)[1-F(y)] \quad \text{a.s. (with respect to F)},$$

where y is such that $0 < F(y) < 1$.

We first show that (34) in fact is valid for all y, for which $0 < F(y) < 1$. Indeed, if $a < b$ are two real numbers such that $0 < F(a) = F(b) < 1$, then

$$\int_a^{+\infty} x\,dF(x) = \int_b^{+\infty} x\,dF(x)$$

but the right hand side of (34) increases from $(c+a)[1-F(a)]$ to $(c+b)[1-F(a)]$. Hence, if a and b have a neighborhood in which $F(x)$ increases then, by the continuity of $F(x)$, (34) fails in some neighborhoods of a and b, although these neighborhoods have positive F-measure. This contradiction leads to the property that $F(x)$ is strictly increasing for all x for which $0 < F(x) < 1$. But then, if (34) holds for almost all y then it holds for all y because $F(y)$ is assumed to be continuous (y is again restricted to $0 < F(y) < 1$).

The completion of the proof is now routine. If we integrate by parts, we get (see Lemma 1.2.1)

$$\int_y^{+\infty} x\,dF(x) = y[1-F(y)] + \int_y^{+\infty} [1-F(x)]dx \ .$$

Hence (34) becomes

$$\int_y^{+\infty} [1-F(x)]dx = c[1-F(y)]$$

whenever $0 < F(y) < 1$. Therefore

$$(35) \quad F(y) = 1 - \exp[-\frac{1}{c}(y-B)] \ ,$$

where B is an arbitrary constant and y is such that $0 < F(y) < 1$. Since $F(y)$ is continuous by assumption, (35) is valid for all $y \geq B$, which was to be proved. □

The above result, and essentially the proof itself, is due to T. Ferguson (1967). Ferguson actually obtains the same conclusion when the sample size is not restricted to $n = 2$, which modifies $F^*(x)$ of (33) slightly, but the argument remains unchanged (the case

$$E(X_{m+1:n}-X_{m:n}|X_{m:n} = y) = \text{constant}$$

can easily be reduced to

$$E(X_{2:n}-X_{1:n}|X_{1:n} = y) = \text{constant}$$

by an appeal to Theorem 3.1.3). Implicitly, some results of Ferguson are contained, under some additional assumptions, in the works by G.S. Rogers (1963) and M.S. Srivastava (1967).

An interesting observation by M. I. Beg and S.N.U.A. Kirmani (1974) is that the following result can be reduced to Theorem 3.4.5.

Theorem 3.4.6. Let $F(x)$ be continuous and assume that $E(X_1)$ is finite. If $\overline{X}$ denotes the arithmetical mean $(1/n)(X_1+\ldots+X_n)$ and if $E(\overline{X}-y|X_{1:n} = y)$ is constant almost surely (with respect to F), then $F(x)$ is exponential.

Proof: We first calculate the conditional distribution of X_j, given $X_{1:n} = y$. We easily get either by direct calculations, or from Theorem 3.1.3, that

$$P(X_j<x|X_{1:n}= y) = \frac{1}{n} + \frac{n-1}{n}\frac{F(x)-F(y)}{1-F(y)} \quad \text{if } x > y ,$$

while this conditional distribution is zero for $x \le y$. Hence,

$$(36) \quad E(X_j|X_{1:n}= y) = \frac{y}{n} + \frac{n-1}{n[1-F(y)]}\int_y^{+\infty} x dF(x) \quad \text{a.s. (F)} .$$

Now observe that the right hand side in (36) does not depend on j. Thus, if we denote the conditional expectation

$$(37) \quad E(X_j|X_{1:n}= y) = g_n(y) ,$$

then $E(\overline{X}|X_{1:n}= y) = \frac{1}{n}\sum_{j=1}^{n}E(X_j|X_{1:n}= y)$ also equals $g_n(y)$. However, $E(\overline{X}|X_{1:n}= y) = y + c_n$, hence

$$(38) \quad g_n(y) = y + c_n , \quad \text{a.s. (F)} .$$

Furthermore, notice that, in view of (36),

$$\frac{1}{n-1} E(nX_j-y|X_{1:n}= y) = \frac{1}{n-1} E(nX_j-X_{1:n}|X_{1:n}= y)$$

does not depend on n. Thus its value is the same for all n. In particular, for $n = 2$ we get from (37) and (38)

$$E(2X_j-X_{1:2}|X_{1:2}= y) = 2E(X_j|X_{1:2}= y) - y = y + c_2 .$$

On the other hand, since $E(X_1|X_{1:2}= y) = E(X_2|X_{1:2}= y)$ almost surely,

$$E(2X_j-X_{1:2}|X_{1:2}= y) = E(X_1+X_2-X_{1:2}|X_{1:2}= y)$$

$$= E(X_{1:2}+X_{2:2}-X_{1:2}|X_{1:2}= y) = E(X_{2:2}|X_{1:2}= y) .$$

Conbining these two expressions, we obtain

$$E(X_{2:2}|X_{1:2}= y) = y + c_2 \quad \text{a.s. (F)} ,$$

which is exactly the assumption of Theorem 3.4.5. We thus get the conclusion of Theorem 3.4.6 from Theorem 3.4.5. □

Theorem 3.4.6 was first proved in a dissertation by Y.H. Wang (1971) and independently by A.C. Dallas (1973), whose direct proofs are also quite simple. In a recent paper, R.C. Srivastava and Y.H. Wang (1978) considered the problem of characterizing distributions by assumptions on the regression $E(Z_k|X_{k:n}=y)$ where

$$Z_k = \sum_{j=k+1}^{n} (X_{j:n} - X_{k:n}) .$$

Notice that

$$E(\bar{X}-y|X_{1:n}=y) = \frac{1}{n} E(\sum_{j=1}^{n} X_{j:n} - nX_{1:n}|X_{1:n}=y) = \frac{1}{n} E(Z_1|X_{1:n}=y) .$$

Consequently, the investigation by Srivastava and Wang is related to Theorem 3.4.6. One way of reducing the case of general k to k = 1 would be to apply Theorem 3.1.3. Instead of doing this, however, Srivastava and Wang develop a direct method of proof, which will probably be fruitful in other investigations related to functions of order statistics.

All results in terms of regression can easily be extended to linear rather than constant regressions. Such an extension will lead, e.g. in (34) to a coefficient of y different from one, but no other change is required in any of the preceding arguments. Depending on this coefficient, of course, we get different families of distributions as solutions to the corresponding regression problem. The details of these calculations are left to the reader.

CHAPTER 4

CHARACTERIZATIONS OF THE POISSON PROCESS

4.1. Introduction.

The simplest way to define the Poisson process is as follows. Let $X_1,X_2,\ldots$ be independent and identically distributed nonnegative random variables. Then the sequence $S_1,S_2,\ldots$ of points, where

$$S_n = X_1 + X_2 +\ldots+ X_n ,$$

is called a *renewal* process. In the case when the common distribution function of the X_j is $F(x) = 1-e^{-bx}$, $b > 0$, $x \geq 0$, then the process $\{S_n, n\geq 1\}$ is called a *Poisson* process. Emphasizing another viewpoint, the renewal process $\{S_n, n\geq 1\}$ is a sequence of random variables $0 \leq S_1 \leq S_2 \leq \ldots$, for which the intervals S_j-S_{j-1}, $j \geq 1$, with $S_0 = 0$, are independent and identically distributed random variables. Throughout this chapter, we shall assume that the common distribution function $F(x)$ of the intervals S_j-S_{j-1} is continuous. Hence, $S_j>S_{j-1}$ with probability one. We shall refer to the S_j as *points* of the process and to $F(x)$ as the *interval distribution* of the process. The notation $\{S_n, n\geq 1; F(x)\}$ will always mean a renewal process whose points are S_n, $n\geq 1$, and whose interval distribution is $F(x)$. Furthermore, S_0 is always zero.

A more general concept than that of a renewal process is a *point* process. Here we adopt the simplest definition by calling any set $\{\tau_s, s\in T\}$ of nonnegative random variables τ_s a point process. Hence, if $T = \{0,1,2,\ldots\}$ and the differences $\tau_s-\tau_{s-1}$, $s\geq 1$, $\tau_0 = 0$, are independent and identically distributed random variables then the point process $\{\tau_s, s\in T\}$ is a renewal process.

In a point process, the variables τ_s will also be referred to as the points of the process. We always denote the points by τ_s if we speak of a general point process. However, if the process is known to be renewal then the previous notation $\{S_n, n\geq 1; F(x)\}$ will be used.

For a given point process, $N(t)$ is defined as the (random) number of points in the interval $(0,t)$. For a renewal process, $N(t)$ is finite with probability one. On the other hand, in all cases of our investigation, the conditions on a point process will guarantee the finiteness (with probability one) of $N(t)$. Hence, we can speak of $N(t_2) - N(t_1)$, which is the number of points of the process under consideration in the interval (t_1,t_2), $t_1 < t_2$. Since the distributions of the points τ_s are always assumed to be continuous, endpoints of intervals can be ignored or included at will.

Let us return to renewal processes. The reason a Poisson process is named

"Poisson" is that

$$(1) \quad P(N(t)=k) = \frac{(bt)^k e^{-bt}}{k!}, \quad k = 0,1,2,\ldots,$$

where $b > 0$ is the parameter of the interval distribution $F(x) = 1-e^{-bx}$, $x \geq 0$. We prove (1) by the following simple argument. Notice that

$$(2) \quad P(N(t)=0) = P(S_1 \geq t) = 1-F(t) = e^{-bt},$$

which is formula (1) with $k = 0$. Now, since

$$P(N(t)=1) = P(S_1<t,\ S_2 \geq t),$$

the continuous version of the total probability rule yields

$$(3) \quad P(N(t)=1) = \int_0^{+\infty} P(S_1<t,\ S_2\geq t \mid S_1=x) be^{-bx} dx$$

$$= \int_0^t P(S_2 \geq t \mid S_1=x) be^{-bx} dx = \int_0^t P(X_2 \geq t-x) be^{-bx} dx$$

$$= b\int_0^t e^{-b(t-x)} e^{-bx} dx = bt\, e^{-bt}.$$

Similarly, since

$$P(N(t)=k) = P(S_k<t,\ S_{k+1}\geq t),\ k \geq 1,$$

we have

$$(4) \quad P(N(t)=k) = \int_0^{+\infty} P(S_k<t,\ S_{k+1}\geq t \mid S_k=x) f_k(x) dx$$

$$= \int_0^t P(S_{k+1}\geq t \mid S_k=x) f_k(x) dx = \int_0^t P(X_{k+1}\geq t-x) f_k(x) dx,$$

where $f_k(x)$ is the density function of S_k. It is, however, well known (and it can easily be obtained by the convolution formula) that if the X_j are exponential variables then $f_k(x)$ is a gamma (Erlang) density. That is,

$$(5) \quad f_k(x) = \frac{b^k x^{k-1} e^{-bx}}{(k-1)!}, \quad x \geq 0,\ k \geq 1,$$

and thus, from (4),

$$P(N(t)=k) = \int_0^t e^{-b(t-x)} \frac{b^k x^{k-1} e^{-bx}}{(k-1)!} dx,$$

which is indeed equal to the value given in (1).

In the next section we shall present a basic characterization theorem for Poisson processes, which theorem can serve as an alternate definition as well. Before turning to the mathematical theory, however, let us list some examples of renewal processes.

At a busy telephone exchange, the successive calls arriving to a given operator can usually be assumed to form a renewal process. Similarly, the successive breakdowns of a new piece of equipment (within a reasonable warranty period); arrivals of buses at a stop; successive time epochs of accidents, and others can serve as typical examples for renewal processes.

Now if a practical model can be assumed to be a renewal process then in order to reduce it to a Poisson process it is required to characterize its interval distribution $F(x)$ to be exponential. This is how the present chapter relates to the others in the present monograph.

There are two ways to apply previous characterization theorems for characterization of Poisson processes within the class of all renewal processes. Namely, if $\{S_n, n\geq 1; F(x)\}$ is a renewal process, then the sequence $X_1 = S_1$ and $X_j = S_j - S_{j-1}$, $j \geq 2$, consists of independent and identically distributed random variables with common distribution function $F(x)$. Hence, through the evolution of a single process, we can form order statistics of the variables $X_1, X_2, \ldots, X_N$, as well as use other properties of this sequence which lead to a (unique) family of distributions for $F(x)$. Another method of characterizing $F(x)$ is to observe the process $\{S_n, n\geq 1; F(x)\}$ several times and to use these observations to determine $F(x)$. In this second method we do not have to follow up the whole process; it suffices to make observations on the first occurrence S_1 only. Let us illustrate these two alternative methods by an example for each case.

Example 1. Let $\{S_n, n\geq 1; F(x)\}$ be a renewal process. Hence, $X_1 = S_1$ and $X_j = S_j - S_{j-1}$, $j \geq 1$, are independent random variables with common distribution function $F(x)$. Let us combine the variables X_{2j+1} and X_{2j+2}, $j \geq 0$, and put

$$W_j = \min(X_{2j+1}, X_{2j+2}) \quad \text{and} \quad U_j = |X_{2j+1} - X_{2j+2}| \; .$$

Now, through the evolution of the process $\{S_n, n\geq 1; F(x)\}$, we can observe (W_0, U_0), $(W_1, U_1), \ldots, (W_N, U_N)$. If, by a statistical method, we can conclude that the components of (W_j, U_j) are independent, then Theorem 3.3.2 implies that $F(x)$ is exponential and thus that the process in question is Poisson.

Example 2. Again let $\{S_n, n\geq 1; F(x)\}$ be a renewal process. Instead of the evolution of the process, we now make independent observations $S_{1,1}, S_{1,2}, \ldots$ on the time S_1 to the first occurrence of an event in the process $\{S_n\}$. Now, if for one $k \geq 2$, k times the smallest $W_{1,k}$ of $S_{1,1}, S_{1,2}, \ldots, S_{1,k}$ is distributed as the $S_{1,j}$, that is, if the distribution function of $kW_{1,k}$ is $F(x)$, and if $F(x)/x \to b > 0$

as $x \to 0^+$, then Theorem 3.2.2 implies that the process in question is Poisson.

While these two methods provide a wide variety of tools to decide whether a renewal process is Poisson, our aim here is different from listing these possibilities as corollaries to the theory developed in the previous chapters. We shall rather investigate properties which involve both the points S_n, $n \geq 1$, and the number $N(t)$ of points in a fixed interval $(0,t)$. Thus new questions arise which cannot be formulated in terms of a single sequence or a single distribution. On the other hand, we shall try to relate all new characterizations to be obtained to previous ones whenever it is possible.

4.2. Basic properties of Poisson processes.

Starting with relation (1), we immediately obtain the following properties of a Poisson process:

(i) $\lim_{t=0} \frac{1-P(N(t)=0)}{t} = b > 0$;

(ii) P (no point is in (t_1,t_2)) depends on the length t_2-t_1 of the interval only;

(iii) the events {no point is in (t_j,t_{j+1})}, $t_j<t_{j+1}$, $j\geq 1$, are independent if the intervals (t_j,t_{j+1}), $j\geq 1$, are disjoint;

and

(iv) $\lim_{t=0} \frac{P(N(t)=1)}{t} = b > 0$.

Indeed, (iv) is obtained by a straight substitution into (1), while (i) follows from (1) upon applying a finite Taylor expansion of $1-e^{-bt}$. For (ii), notice that

$$(6) \quad P(\text{no point is in } (t_1,t_2)) = P(N(t_2) - N(t_1) = 0)$$

$$= \sum_{k=0}^{+\infty} P(N(t_1) = k,\ N(t_2) = k) = \sum_{k=0}^{+\infty} P(S_k<t_1,\ S_{k+1}\geq t_2) ,$$

where $S_0 = 0$. Applying again the argument used in (4), we get, for $k \geq 1$,

$$P(S_k<t_1,\ S_{k+1}\geq t_2) = \int_0^{t_1} P(S_{k+1}\geq t_2 | S_k=x) f_k(x)dx$$

$$= \int_0^{t_1} P(X_{k+1}\geq t_2-x) f_k(x)dx = e^{-bt_2} \frac{b^k t_1^k}{k!} .$$

This formula is evidently true for $k = 0$ as well. Hence, on account of (6),

$$P(\text{no point is in } (t_1,t_2)) = \exp[-b(t_2-t_1)] ,$$

which shows that this expression is a function of t_2-t_1 only as claimed in (ii).

The proof of (iii) is similar to the preceding argument. We demonstrate it for two disjoint intervals (t_1,t_2) and (t_3,t_4), where $t_1 < t_2 \le t_3 < t_4$. Let A and B denote the events that no points of the process fall into the intervals (t_1,t_2) and (t_3,t_4), respectively. Then AB is the event that, for some integers $0 \le k \le m$, $\{S_k<t_1,\ S_{k+1}\ge t_2$ and $S_m<t_3,\ S_{m+1}\ge t_4\}$. If this latter event is denoted by $C_{k,m}$, then we have

$$(7)\qquad P(AB) = \sum_{k=0}^{+\infty}\ \sum_{m=k}^{+\infty} P(C_{k,m}) \ .$$

We now distinguish two cases. First we evaluate $P(C_{m,m})$. Applying the method utilized for the derivation of (4), we get

$$(8)\qquad P(C_{m,m}) = \int_0^{t_1} P(S_{m+1}\ge t_4 | S_m = x) f_m(x)dx$$

$$= \int_0^{t_1} P(X_{m+1}\ge t_4-x) f_m(x)dx = e^{-bt_4}\frac{bt_1^m}{m!} \ .$$

We now turn to $P(C_{k,m})$ with $k < m$. Since, in this case, $S_k < t_1$, $S_{k+1} \ge t_2$ but $S_{k+1} \le S_m < t_3$, and $S_{m+1} \ge t_4$, we have

$$(9)\qquad P(C_{k,m}) = \int_0^{t_1}\int_{t_2}^{t_3} P(S_{k+1}\ge t_2,\ S_{m+1}\ge t_4 | S_k = x,\ S_m = y) g_{k,m}(x,y)dydx$$

$$= \int_0^{t_1}\int_{t_2}^{t_3} P(t_2-x\le X_{k+1}\le y-x, X_{m+1}\ge t_4-y | S_k = x,\ S_m = y) g_{k,m}(x,y)dydx \ ,$$

where $g_{k,m}(x,y)$ is the joint density function of the vector (S_k,S_m), $k < m$. Since

$$S_m = S_k + (X_{k+1}+\ldots+X_m) = S_k + S^*_{m-k}\ , \text{ say,}$$

$g_{k,m}(x,y)$ can easily be evaluated by observing that S_k and S^*_{m-k} are independent with density functions $f_k(x)$ and $f_{m-k}(z)$, respectively, where $f_k(x)$ is the function defined at (5). Furthermore, $\{S_k = x,\ S_m = y\} = \{S_k = x,\ S^*_{m-k} = y-x\}$. Therefore

$$(9a)\qquad g_{k,m}(x,y) = f_k(x) f_{m-k}(y-x)\ ,\ 0 \le x \le y,\ 1 \le k < m.$$

The evaluation of $P(AB)$ in (7) can now be easily completed as follows. Upon observing the special relation of the condition $\{S_k = x,\ S_m = y\}$ in (9) to X_{k+1} and X_{m+1}, we first simplify the integrand in (9) by an additional conditioning on $\{X_{k+1} = z\}$. We then substitute these integrals into (7), where interchanging summation and integration and utilizing (9a) and (5) yield

$$P(AB) = \exp[-b(t_2-t_1)]\exp[-b(t_4-t_3)] \ .$$

This shows the independence of A and B in view of the basic property (ii), established earlier.

The proof of (iii) for more than two disjoint intervals (t_j,t_{j+1}), $j \geq 1$, is quite similar and we shall omit the computational details.

Let us now return to the basic properties (i) - (iv). Notice that, in view of the obvious equation

$$(10) \quad P(N(t) = 0) + P(N(t) = 1) + P(N(t) \geq 2) = 1 \ ,$$

properties (i) and (iv) imply

$$(v) \quad \lim_{t=0} \frac{P(N(t)\geq 2)}{t} = 0 \ .$$

In fact, a direct consequence of (1) is that

$$P(N(t)\geq 2) = \sum_{k=2}^{+\infty} \frac{(bt)^k e^{-bt}}{k!} \leq e^{-bt}(bt)^2 \sum_{k=2}^{+\infty} \frac{(bt)^{k-2}}{(k-2)!} = (bt)^2 \ .$$

For the sake of future reference, we include this last estimate in the following

<u>Theorem 4.2.1</u>. The Poisson process $\{S_n, n\geq 1; F(x)\}$, where $F(x) = 1 - e^{-bx}$, $x \geq 0$, satisfies the basic properties (i) - (iv). Furthermore, the following inequality

$$(vi) \quad P(N(t)\geq 2) \leq (bt)^2 \ ,$$

holds.

As we shall see later, properties (i) - (iv) and (vi) characterize the Poisson process among point processes. Notice the fact that these properties do not refer to the fact that the points of the process in question form a renewal process: each of the properties (i) - (vi) is formulated in terms of the distribution of N(t) (with varying t). It is, of course, evident that some of the properties (i) - (vi) follow from the others. For example, (v) is a consequence of (vi), or of the combination of (i) and (iv). Our aim with stating all of these properties is to emphasize the most basic ones pertaining to a Poisson process. In fact, several text books start with these properties to define a Poisson process. However, we wanted to relate Poisson processes to the exponential distribution from the very start, which warranted our particular definition.

Before proving that properties (i) - (vi) characterize the Poisson process among point processes $\{\tau_j\}$ (with some minor restrictions which would guarantee that $P(\tau_j = \tau_k) = 0$ for all $j \neq k$), we deduce a consequence of the basic properties

(i) - (iv).

To simplify expressions, we introduce an additional notation. Let A be the union of a finite number of disjoint finite intervals (t_j, t_{j+1}), $1 \le j \le M$. We define N(A) to be the number of points of the process in question which fall into the set A. Hence

$$N(A) = \sum_{j=1}^{M} [N(t_{j+1}) - N(t_j)] .$$

We always assume that the process $\{\tau_j\}$ is such that $P(\tau_j = t) = 0$ for any t and $P(\tau_j = \tau_k) = 0$ for all $j \neq k$, hence the endpoints of intervals may be omitted or included at will without affecting the results. For renewal processes, we can achieve this by assuming that the interval function F(x) is continuous, but for our very general definition of a point process, this requires more than the continuity of the distribution function at each point (the reader will find examples in Chapter 5 where a vector (X,Y) is such that $X \ge 0$, $Y \ge 0$, each component has an exponential distribution, but $P(X=Y) > 0$).

We shall also express $P(N(A) = 0)$ for the sets A introduced above

$$(11) \quad P(N(A) = 0) = e^{-m(A)} ,$$

whenever the left hand side is positive. Although (11) is just a change in notation, it helps us to formulate very simple conditions for $P(\tau_j = \tau_k) = 0$ $(j \neq k)$ in the case of processes which satisfy property (iii). Namely, under (iii), m(A) is a non-negative additive set function on the finite unions of finite intervals. Therefore, the simple additional requirement that m(A) should be nonatomic is equivalent to our conditions on the sequence $\{\tau_j\}$.

We now prove the following general result.

Theorem 4.2.2. Let m(A) in (11) be a nonatomic, nonnegative and additive set function on the finite unions of finite intervals. Assume that the point process $\{\tau_j\}$ is such that, for finite intervals I,

$$(12) \quad P(N(I) \ge 2) \le b|I|^2 ,$$

where $b > 0$ is some fixed constant and $|I|$ signifies the length (or Lebesgue measure) of the interval I. Then, if $I_j = (t_j, t_{j+1})$, $t_j < t_{j+1}$, $1 \le j \le M$, are disjoint intervals, the random variables $N(I_j)$, $1 \le j \le M$, are independent.

Proof: The fact that m(A) in (11) is an additive set function implies that the events $\{N(I_j) = 0\}$, $1 \le j \le M$, are independent. Indeed, if

$$A = \bigcup_{j=1}^{M} I_j ,$$

then

$$m(A) = m(I_1) + m(I_2) + \dots + m(I_M) ,$$

and, thus, in view of (11),

$$P(N(A) = 0) = P(N(I_j) = 0,\ 1 \le j \le M) = e^{-m(A)}$$
$$= \exp[-\sum_{j=1}^{M} m(I_j)] = \prod_{j=1}^{M} e^{-m(I_j)} = \prod_{j=1}^{M} P(N(I_j) = 0) .$$

Since the above argument is evidently valid for any sub-collection of the intervals I_j, we have shown the independence of the events $\{N(I_j) = 0\}$, $1 \le j \le M$.

We now prove the independence of the random variables $N(I_j)$, $1 \le j \le M$. In other words, we want to prove that, for arbitrary nonnegative integers k_j, $1 \le j \le M$, the difference

$$(13) \quad P(N(I_j) = k_j,\ 1 \le j \le M) - \prod_{j=1}^{M} P(N(I_j) = k_j)$$

equals zero. Let us divide the intervals I_j into r equal parts, $I_{j,k}$, say, where now $1 \le j \le M$ and $1 \le k \le r$. Let $e_{j,k}$ be the indicator variable of the event $\{N(I_{j,k}) = 0\}$, i.e., $e_{j,k} = 1$ or 0 according to whether $N(I_{j,k}) = 0$ or $N(I_{j,k}) \ge 1$. Arguing as above with the events $\{N(I_j) = 0\}$, we get that the variables $e_{j,k}$ are independent for all j and k. But then so are the random variables

$$E_j = \sum_{k=1}^{r} (1-e_{j,k}) , \quad 1 \le j \le M .$$

Therefore, if we replace $N(I_j)$ by E_j in (13), then the value of the difference similar to the one in (13) is zero, i.e.,

$$(14) \quad P(E_j = k_j,\ 1 \le j \le M) - \prod_{j=1}^{M} P(E_j = k_j) = 0 .$$

However, since

$$P(N(I_j) \ne E_j) \le \sum_{k=1}^{r} P(N(I_{j,k}) \ge 2) ,$$

we have from (12) that for any fixed j,

$$(15) \quad |P(N(I_j) = k_j) - P(E_j = k_j)| \le P(N(I_j) \ne E_j)$$
$$\le b \sum_{k=1}^{r} |I_{j,k}|^2 = b \sum_{k=1}^{r} \frac{|I_j|^2}{r^2} = \frac{b|I_j|^2}{r} .$$

Similarly,

$$(16)\quad |P(N(I_j) = k_j,\ 1 \le j \le M) - P(E_j = k_j,\ 1 \le j \le M)| \le \frac{b}{r} \sum_{j=1}^{M} |I_j|^2 .$$

Since, b, M and the lengths $|I_j|$ are fixed, relations (14) - (16) imply that the absolute value of the difference in (13) is smaller than c/r, where $c > 0$ is a suitable constant and r is an arbitrary positive integer. Letting $r \to +\infty$ yields that the difference (13) is zero, which means the independence of the random variables $N(I_j)$, $1 \le j \le M$. The proof is completed. □

The meaning of Theorem 4.2.2 is that among the basic properties (i) - (vi), (iii) can freely be replaced by the seemingly stronger condition

(iii*) the random variables $N(I_j)$ are independent, where $I_j = (t_j, t_{j+1})$, $t_j < t_{j+1}$, $1 \le j \le M$, and the intervals I_j are nonoverlapping.

The usual definition of a Poisson process in the literature is to assume that it is a point process satisfying the basic properties (i), (ii*), (iii*) and (iv), where (ii*) is the following modified version of (ii):

(ii*) the distribution of N(I), where I is an interval, depends on the length of the interval only.

We shall now prove that such a definition is equivalent to our definition given in the introduction. Other theorems of the present section will be devoted to the consequences of other combinations of the basic properties (i) - (vi).

<u>Theorem 4.2.3</u>. Let $\{\tau_s\}$ be a point process on the nonnegative real line. Assume that the random variables $N(I_j)$, where the I_j are finite intervals, satisfy conditions (i), (ii*), (iii*) and (iv), where N(t) denotes N(I) for $I = (0,t)$. Then the points $\{\tau_s\}$ can almost surely be arranged into an increasing sequence $\{S_j\}$ and the process $\{S_j,\ j \ge 1\}$ is Poisson.

<u>Proof</u>: We shall first prove that, for any fixed $t > 0$,

$$(17)\quad P(N(t) = k) = \frac{(bt)^k e^{-bt}}{k!}, \quad k = 0,1,2,\ldots,$$

where b is given in (i). (This was established in (1) for Poisson processes.) Let us put

$$P_k(t) = P(N(t) = k),\ k = 0,1,\ldots .$$

We shall prove (17) by induction over k. First let $k = 0$. Then, for any $t \ge 0$ and $s \ge 0$, $\{N(t+s) = 0\}$ means that $N(t) = 0$ and that there are no τ's in $I_1 = (t, t+s)$.

Thus, in view of the assumptions (ii*) and (iii*),

$$P_0(t+s) = P_0(t)P_0(s) .$$

Also, $P_0(t)$ is monotonic. Since, in view of condition (i), $P_0(t)$ is not constant in t, Theorem 1.3.1 implies that

$$P_0(t) = e^{-at}, \; t \geq 0 ,$$

where $a > 0$ is a suitable constant. One more appeal to (i) now yields that $a = b$ and we have thus verified (17) for $k = 0$.

Turning now to $P_1(t)$, we calculate $P_1(t+s)$ as above by splitting the interval $(0,t+s)$ into the union of $(0,t)$ and $I_1 = (t, t+s)$. Since, by (iii*), $N(t)$ and $N(I_1)$ are independent and since

$$\{N(t+s) = 1\} = \{N(t) = 0, N(I_1) = 1\} \cup \{N(t) = 1, N(I_1) = 0\}$$

we have

$$P_1(t+s) = P_0(t)P(N(I_1) = 1) + P_1(t)P(N(I_1) = 0) .$$

If we now apply (ii*) and (17) with $k = 0$, we get from the last equation

$$P_1(t+s) = P_0(t)P_1(s) + P_1(t)P_0(s)$$

$$= P_1(s)e^{-bt} + P_1(t)e^{-bs} .$$

Denote $g_1(u) = P_1(u)e^{bu}$. The above equation then becomes

$$g_1(t+s) = g_1(s) + g_1(t) ,$$

which can be rewritten as

$$\frac{g_1(t+s)-g_1(s)}{t} = \frac{g_1(t)}{t} = \frac{P_1(t)}{t} e^{bt} .$$

On account of (iv), the right hand side converges to $b > 0$ as $t \to 0$. Hence, so does the left hand side, which means that $g_1(s)$ is differentiable and

$$g_1'(s) = b .$$

Consequently,

$$g_1(s) = P_1(s)e^{bs} = bs + c ,$$

where c is a constant. Applying (iv) again, we get $c = 0$, and thus (17) is verified for $k = 1$.

Assume now that (17) is valid for some k and consider $P_{k+1}(t)$. Arguing as before, we obtain

$$(18)\quad P_{k+1}(t+s) = P_{k+1}(t)P_0(s) + P_k(t)P_1(s) + O(P(N(I_1) \geq 2)) ,$$

where $I_1 = (t,t+s)$, and thus, by (i), (ii*) and (iv),

$$P(N(I_1) \geq 2) = P(N(s) \geq 2) = 1 - P(N(s) = 0) - P(N(s) = 1)$$
$$= o(s) \text{ , as } s \to 0.$$

Consequently, if we rearrange (18) as

$$\frac{P_{k+1}(t+s) - P_{k+1}(t)}{s} = P_{k+1}(t)\,\frac{P_0(s)-1}{s} + P_k(t)\,\frac{P_1(s)}{s} + o(s) ,$$

we obtain, on account of (i) and (iv), that $P_{k+1}(t)$ is differentiable and moreover,

$$(19)\quad P'_{k+1}(t) = -bP_{k+1}(t) + bP_k(t) .$$

Applying the assumption of induction to $P_k(t)$ and utilizing the fact that $P_{k+1}(t) \to 0$ as $t \to 0$, which is an immediate consequence of (i), we obtain from (19) that (17) is valid for k+1. Hence, (17) is established.

It now follows from (17) that the points $\{\tau_s\}$ can be rearranged into an increasing sequence $S_1 < S_2 <\ldots$ with probability one. In fact, we have from (17) that, with probability one, any finite interval can contain only a finite number of τ-points. Therefore, the τ-points can be arranged into a single increasing sequence by first considering the points in the interval (0,1), then those in (1,2), (2,3), and so on.

In order to complete the proof, we now show that the sequence $0 < S_1 < S_2 <\ldots$ forms a Poisson process. In other words, we have to prove that the random intervals $S_j - S_{j-1}$, $j \geq 1$, $S_0 = 0$, are independent with common distribution function $F(x) = 1-e^{-bx}$, $x \geq 0$. We shall prove this again by induction. Since

$$P(S_1 \geq x) = P(N(x) = 0) ,$$

we get from (17) that

$$P(S_1 < x) = 1 - e^{-bx}, \; x \geq 0 .$$

Next, we calculate the joint distribution of S_1 and S_2-S_1, by appealing to the continuous version of the total probability rule. We get

$$P(S_1<x,\ S_2-S_1\geq y) = \int_0^{+\infty} P(S_1<x,\ S_2-S_1\geq y \mid S_1 = z)be^{-bz}dz$$

$$= \int_0^{x} P(S_2\geq z+y \mid S_1 = z)be^{-bz}dz \ .$$

However, given $S_1 = z$, the inequality $S_2 \geq z+y$ means that the interval $(z,\ z+y)$ contains no points of the original process. Hence, the probability $P(S_2\geq z+y|S_1 = z)$, on account of (ii*), is equal to $P(N(y) = 0)$, which, by (17) again, equals e^{-by}. Consequently

$$P(S_1<x,\ S_2-S_1\geq y) = be^{-y}\int_0^{x} e^{-bz}dz = e^{-by}(1-e^{-bx}) \ .$$

Hence, we have shown that S_1 and S_2-S_1 are independent and identically distributed. Assume now that $S_1,\ S_2-S_1,\ldots,S_k-S_{k-1}$ are independent with common distribution function $F(x) = 1-e^{-bx}$, $x \geq 0$. Then, putting $d_j = S_j-S_{j-1}$, $j \geq 1$, where $S_0 = 0$,

$$P(d_1<x_1,\ d_2<x_2,\ldots,d_k<x_k,\ d_{k+1}\geq y)$$

$$= \int_0^{+\infty} \cdots \int_0^{+\infty} P(d_j<x_j,\ 1\leq j\leq k,\ d_{k+1}\geq y \mid d_j = z_j,\ 1\leq j\leq k)h_k(z_1,\ldots,z_k)dz_1\ldots dz_k \ ,$$

where

$$h_k(z_1,z_2,\ldots,z_k) = b^k\exp[-b(z_1+z_2+\ldots+z_k)] \ ,$$

the joint density of $(d_1,d_2,\ldots,d_k)$. However, if $z_j < x_j$, $1 \leq j \leq k$, then the conditional probability appearing in the integral above reduces once more to the probability that the interval $(z_1+z_2+\ldots+z_k,\ z_1+z_2+\ldots+z_k+y)$ contains no points of the process. Therefore, we obtain from (ii*) and (17) that this probability equals e^{-by}. On the other hand, if $z_j \geq x_j$ for some j, then this same conditional probability is zero. Consequently, the last integral yields

$$P(d_j<x_j,\ 1\leq j\leq k,\ d_k\geq y) = e^{-by}F(x_1)F(x_2)\ldots F(x_k) \ ,$$

where $F(x) = 1-e^{-bx}$, $x \geq 0$. This completes the proof. □

The results obtained so far in this section can be summarized into the following

Corollary 4.2.1. Let $\{\tau_s,\ s\epsilon T\}$ be a point process, where T is an arbitrary set of real numbers. Let A be the union of a finite number of finite intervals of the nonnegative real line. Assume that $P(N(A) = 0) > 0$ for A's of positive Lebesgue measure, and define $m(A)$ by the equation

$$P(N(A) = 0) = e^{-m(A)} .$$

Let $m(A)$ be a nonnegative, nonatomic, additive set function. Then the process $\{\tau_s\}$ is a Poisson process if, and only if, properties (ii*) and (vi) hold.

Proof: Let us first observe that all the assumptions of the corollary are valid for a Poisson process. Namely, on account of (ii) and (17), $m(A)/b$ equals the Lebesgue measure of the set A, where b is the parameter of the interval distribution $F(x)$. Hence, $m(A)$ is indeed nonnegative, nonatomic and additive. In addition, (ii*) can be verified by the same simple calculation as the one applied at (6). Finally, it was shown in Theorem 4.2.1 that (vi) is valid for a Poisson process.

Conversely, if $m(A)$ is an additive set function, then property (iii) evidently holds. However, property (ii), which is a special case of our assumption (ii*), and property (iii) imply that $m(A)$ is proportional to Lebesgue measure. Indeed, if we set $I_1 = (0,t)$ and $I_2 = (t, t+s)$, where $t > 0$, $s > 0$, then by (ii)

$$P(N(I_2) = 0) = P(N(s) = 0) = e^{-m(s)} ,$$

where, for intervals $I(x) = (0,x)$, we use the notations $m(x) = m[I(x)]$ and $N(x) = N[I(x)]$. On the other hand, by (iii),

$$\begin{aligned} e^{-m(I_1 \cup I_2)} &= P(N(I_1 \cup I_2) = 0) = P(N(I_1) = 0,\ N(I_2) = 0) \\ &= P(N(I_1) = 0)P(N(I_2) = 0) = e^{-m(I_1)}e^{-m(I_2)} , \end{aligned}$$

and thus

$$(20) \quad m(t+s) = m(t) + m(s) , \quad t,s > 0.$$

This is a variant of the lack of memory equation

$$G(t+s) = G(t)G(s),\ t,\ s \geq 0 ,$$

through the relation $m(x) = -\log G(x)$. Since, by definition, $m(x)$ is nondecreasing, Theorem 1.3.1 implies that $m(x) = bx$ for all $x \geq 0$, where $b > 0$ is a constant, because $m(A)$ is nonatomic and is positive for all A of positive Lebesgue measure (and thus $m(x)$ cannot be constant over a finite interval). This specific form of $m(x)$ implies the validity of property (i). Therefore, (iv) also holds in view of our assumption (vi), while Theorem 4.2.2 implies that (iii*) is valid. Hence, Theorem 4.2.3 is applicable, which completes the proof. □

In the preceding argument it was crucial to guarantee the validity of (20) for an arbitrary additive set function $m(A)$. This was achieved by the assumption that (ii) should hold (we assumed the stronger version (ii*), but, for (20), the weaker form (ii) would have sufficed). The question therefore arises whether condition (ii) is essential for the conclusion of Corollary 4.2.1 itself (i.e., perhaps with another proof we could have avoided it). We shall see from the following theorem, which is due to A. Rényi (1967), that the answer to this question is yes; although the distribution of N(A) remains Poisson without assuming (ii) or (ii*), the point process $\{\tau_s\}$ is not Poisson in the sense of our definition unless (ii) is valid.

Theorem 4.2.4. For the point process $\{\tau_s, s\epsilon T\}$ let the set function m(A) be defined as in Corollary 4.2.1. As before, let m(A) be nonatomic and additive. If, for finite intervals I,

$$(21)\quad P(N(I) \geq 2) \leq bm^2(I) ,$$

where $b > 0$ is a constant, then

$$(22)\quad P(N(I) = k) = \frac{m^k(I)e^{-m(I)}}{k!} , \quad k = 0,1,2,\ldots .$$

Remark 1. The relation of (21) to (12) or to property (vi) is evident. If m(I) is proportional to Lebesgue measure then (21) reduces to (12) and if, in addition the distribution of N(I) depends only on the length of the interval I, then it further reduces to (vi).

Remark 2. The assumption in (21) can be weakened to the following condition without affecting the conclusion of the theorem: there is a strictly decreasing function $g(x) > 0$ which tends to zero as $x \to 0$ and such that

$$P(N(I) \geq 2) \leq m(I)g[m(I)] .$$

Remark 3. The assumption that m(A) is nonatomic implies that $m(I) \to 0$ as $d \to c$ where I is the interval [c,d). Namely, otherwise c would be an atom of the function m(A). Similarly, we obtain that m(I) is a continuous function of d.

Remark 4. Notice that (22) again implies that $\{\tau_s\}$ can be rearranged into a single nondecreasing sequence $S_1 \leq S_2 \leq \ldots$. However, this sequence is not a Poisson process unless m(A) is proportional to Lebesgue measure (which is equivalent to the validity of (20), which in turn is equivalent to the validity of (ii)). As a matter of fact,

$$P(S_1 < x) = 1 - P(N(x) = 0) = 1 - e^{-m(x)} ,$$

where we again abbreviated m[(0,x)] as m(X). Hence, even if the sequence $S_1 \leq S_2 \leq \ldots$ were a renewal process, its interval distribution is, in general, not exponential. We can also see from Theorem 4.2.4 that, in Corollary 4.2.1, (ii*) can be replaced by the weaker assumption (ii).

Proof of Theorem 4.2.4. Let $I = (c,d)$ and consider the characteristic function

(23) $\phi(u) = E(e^{iuN(I)})$.

Let us divide the interval into M equal subintervals, and put

$$I_j = (c + \frac{(j-1)(d-c)}{M}, c + \frac{j(d-c)}{M}) , j = 1,2,\ldots,M .$$

Hence,

(24) $N(I) = N(I_1) + N(I_2)+\ldots+N(I_M)$.

Now, since $m(A)$ is additive, and since

$$e^{-m(I)} = P(N(I) = 0) = P(N(I_j) = 0, 1 \le j \le M) ,$$

we have

$$P(N(I_j) = 0, 1 \le j \le M) = \exp[-\sum_{j=1}^{M} m(I_j)]$$

$$= P(N(I_1) = 0)P(N(I_2) = 0)\ldots P(N(I_M) = 0) .$$

In other words, the events $\{N(I_j) = 0\}, 1 \le j \le M$, are independent. We can now use the same argument as in the proof of Theorem 4.2.2 to conclude that the independence of the events above and (21) imply that the random variables $N(I_j)$, $1 \le j \le M$, are also independent. Therefore, on account of (23) and (24),

(25) $\phi(u) = \prod_{j=1}^{M} E(e^{iuN(I_j)})$.

However,

$$\phi_j(u) = E\{\exp[iuN(I_j)]\}$$

$$= E(e^{iuN(I_j)}|N(I_j) = 0)P(N(I_j) = 0) + E(e^{iuN(I_j)}|N(I_j) = 1)P(N(I_j) = 1) + R_j$$

$$= P(N(I_j) = 0) + e^{iu}P(N(I_j) = 1) + R_j ,$$

where

$$R_j = E(e^{iuN(I_j)}|N(I_j) \ge 2)P(N(I_j) \ge 2) = O(m^2(I_j)).$$

Hence

(26) $\phi_j(u) = e^{-m(I_j)} + e^{iu}(1-e^{-m(I_j)}) + O(m^2(I_j))$.

As $M \to +\infty$, $m(I_i) \to 0$ (see Remark 3). Therefore, by Taylor's expansion, (26) becomes

$$\phi_j(u) = 1-m(I_j)(1-e^{iu}) + O(m^2(I_j)) ,$$

from which the Taylor expansion of $\log \phi_j(u)$ yields

$$(27) \quad \log \phi_j(u) = m(I_j)(e^{iu}-1) + O(m^2(I_j)) ,$$

where the error term is uniform in j. Applying (27) to the logarithm of (25), we get

$$(28) \quad \log \phi(u) = \sum_{j=1}^{M} m(I_j)(e^{iu}-1) + O(\sum_{j=1}^{M} m^2(I_j))$$

$$= m(I)(e^{iu}-1) + O(\max_{1\le j\le M} m(I_j)) .$$

Notice that the left hand side does not depend on M, hence the limit of the right hand side, as $M \to +\infty$, gives $\log \phi(u)$. As we have noted, however, (Remark 3), $m(I)$ is continuous in the upper end point of I, and thus, Heine's theorem of elementary calculus implies that it is uniformly continuous. Consequently,

$$\max_{1\le j\le M} m(I_j) \to 0 \text{ as } M \to +\infty$$

and thus we get from (28) that

$$\phi(u) = \exp[m(I)(e^{iu}-1)] .$$

This is, however, equivalent to (22) by the inversion formula of characteristic functions. The theorem is established. □

We conclude this section with an example exhibiting the possible general nature of $m(A)$. Let $g(t)$ be a nonnegative continuous function for $t \ge 0$. Let

$$(29) \quad m(A) = \int_A g(t)dt ,$$

where A is always the union of a finite number of finite intervals. Hence, the integral in (29) is a Riemann integral. For the special case $A = (0,x)$, in which case we write $m(x) = m(A)$,

$$(30) \quad m(x) = \int_0^x g(t)dt .$$

Evidently, $m(x)$ uniquely determines $m(A)$. Notice that $g(t) = b > 0$ leads to a Poisson process in Theorem 4.2.4. The function $g(t)$ has the following important meaning under the conditions of Theorem 4.2.4. Let $I = (t, t+s)$ in (22). Then

$$P(N(I) = 0) = \exp[-\int_t^{t+s} g(u)du]$$

and thus

$$\lim_{s=0} \frac{1-P(N(I) = 0)}{s} = g(t) .$$

In addition, again by (22),

$$\lim_{s=0} \frac{P(N(I) = 1)}{s} = g(t) .$$

It is instructive to compare these limit relations with the basic formulas (i) and (iv). Their meanings are that, in short intervals (t, t+s), the probability of exactly one point falling into (t, t+s), is approximately linear in the length of the interval both for a Poisson process and for processes described in Theorem 4.2.4. Hence, what actually distinguishes a Poisson process from a process represented in Theorem 4.2.4 is that this probability does not depend on the location of (t, t+s) for a Poisson process, while the location is important in the latter case. Because of this strong relation between these two types of processes, we shall call a point process satisfying the conditions of Theorem 4.2.4 with m(A) defined in (29), a *time dependent* Poisson process.

4.3. Characterizations based on age and residual life.

Let $\{S_n, n\geq 1; F(x)\}$ be a renewal process. For fixed $t > 0$ we call the random variables

(31) $a(t) = t-S_{N(t)}$

and

(32) $r(t) = S_{N(t)+1}-t$,

the age and the residual life at t, respectively,where, as always, N(t) denotes the number of the points of the process in the interval (0,t). In other words, N(t) is such that

$$S_{N(t)} < t \leq S_{N(t)+1} .$$

Since we always assume F(x) to be continuous, the equality actually has no effect above and thus both a(t) and r(t) are positive random variables (with probability one).

The distributions of r(t) and a(t) take simple forms for a Poisson process. As a matter of fact, the distribution of r(t) is F(x) itself and that of a(t) is $F_t(x)$ which is F(x) for $x < t$ and $F_t(x) = 1$ for $x > t$. We shall develop several characterizations of Poisson processes based on these simple (but nonetheless surprising) properties of $r(t)$ and $a(t)$. The reader will recognize the following consequence of the mentioned properties, which is known in the literature as the *waiting time paradox*. As an example, consider the sequence $0 < S_1 < S_2 < \dots$ as the successive arrival time epochs of buses at a given station. Then the intervals $S_j - S_{j-1}$ are identically distributed with common distribution function $F(x)$, except that interval when I want to board a bus: that special bus I was waiting for comes later after the preceding one as compared with other consecutive buses. Namely, if t represents my arrival time then $S_{N(t)}$ is the time epoch of the bus I missed and $S_{N(t)+1}$ is the one that I will board. Hence, this special time interval

$$S_{N(t)+1} - S_{N(t)} = a(t) + r(t)$$

is longer than a "usual" interval $S_j - S_{j-1}$, since $r(t)$ above contributes as much as $S_j - S_{j-1}$ in the case of "the other buses."

In order to prove this paradox and to deduce several characterizations from it, we evaluate the distribution of $a(t)$ and $r(t)$, respectively, for an arbitrary renewal process. In the calculations we shall need the following function $U(x)$. For the interval distribution function $F(x)$, this is defined as

$$(33)\quad U(x) = \sum_{n=0}^{+\infty} F^{(n)}(x) ,$$

where $F^{(n)}(x)$ denotes the n-fold convolution of $F(x)$. We call $U(x)$ the *renewal function*. Notice the special meaning of $U(x)$. We claim that

$$(34)\quad U(x) = E[N(x)] + 1 .$$

Indeed, since

$$P(N(x) \geq n) = P(S_n < x) = F^{(n)}(x) ,$$

$$E[N(x)] = \sum_{n=1}^{+\infty} nP(N(x) = n) = \sum_{n=1}^{+\infty} P(N(x) \geq n)$$

$$= \sum_{n=1}^{+\infty} F^{(n)}(x) = U(x) - 1 ,$$

as claimed. Relation (34) can be interpreted as saying $U(x)$ represents the expected number of points of the process which fall into the interval $[0,x)$, where $S_0 = 0$ is also counted.

It is instructive to verify (34) for a Poisson process directly. We have seen that the distribution of $N(x)$ is Poisson with parameter bx. Hence, the right hand side of (34) equals $bx + 1$. On the other hand, the left hand side can be computed by the exact formula (see (5))

$$F^{(n)}(x) = \int_0^x \frac{b^n t^{n-1} e^{-bt}}{(n-1)!} dt .$$

We have

$$U(x) = 1 + \int_0^x \sum_{n=1}^{+\infty} \frac{b^n t^{n-1} e^{-bt}}{(n-1)!} dt = 1 + \int_0^x bdt = 1 + bx .$$

We should add that the calculation of $F^{(n)}(x)$, and thus of $U(x)$, can be very difficult even for very simple functions $F(x)$. In most cases, however, calculations can be simplified by turning to Fourier transforms.

It is evident from its definition that $U(x)$ is increasing and $U(0^+) = 1$. Hence, we can speak of integrals with respect to $U(x)$ if we establish that $U(x)$ is finite for all x. This is the content of the next lemma.

Lemma 4.3.1. The renewal function $U(x)$ is finite for all x.

Proof: Recall that $F(x)$ is continuous with $F(0) = 0$. We first prove that, for an arbitrary $y > 0$, there is an integer $M \geq 1$ such that $F^{(M)}(y) < 1$. Indeed, since $F(y)$ is continuous, $F(0) = 0$ and $F(+\infty) = 1$, there is an $x_0 > 0$ such that $0 < F(x_0) < 1$. It then follows that, for any integer $m \geq 1$,

$$F^{(m)}(mx_0) < 1.$$

Namely, $F(x_0) < 1$ means that $P(S_1 \geq x_0) > 0$ and

$$\begin{aligned} F^{(m)}(mx_0) &= P(S_m \geq mx_0) \geq P(S_j - S_{j-1} \geq x_0,\ 1 \leq j \leq m) \\ &= F^m(x_0) > 0 , \end{aligned}$$

where $S_0 = 0$. Hence, for $y > 0$, we can choose any $M \geq 1$ such that $Mx_0 \geq y$, for which it is assured that $F^{(M)}(y) < 1$.

Now, by the convolution formula,

$$\begin{aligned} F^{(n)}(y) &= \int_0^y F^{(m)}(y-z) dF^{(n-m)}(z) \\ &\leq F^{(m)}(y) \int_0^y dF^{(n-m)}(z) = F^{(m)}(y) F^{(n-m)}(y) . \end{aligned}$$

Therefore, if M is such that $F^{(M)}(y) < 1$, then, writing $n = Mk + R$, where $0 \leq R < M$, we have

$$F^{(n)}(y) \leq [F^{(M)}(y)]^k .$$

Hence,

$$U(y) \le M \sum_{k=0}^{+\infty} [F^{(M)}(y)]^k = \frac{M}{1-F^{(M)}(y)} < +\infty ,$$

which completes the proof. □

We can now calculate the distribution of $r(t)$.

Lemma 4.3.2. For a renewal process $\{S_n, n\ge 1; F(x)\}$, the distribution functior of the residual life $r(t)$ at t is given by

$$R_t(x) = P(r(t)<x) = 1 - \int_0^t [1-F(t+x-y)]dU(y) ,$$

where $U(y)$ is the function defined at (33).

Proof: Let us calculate

$$1 - R_t(x) = P(r(t) \ge x) .$$

We decompose this last probability as

$$(35) \quad P(r(t)\ge x) = \sum_{n=0}^{+\infty} P(r(t)\ge x,\ N(t) = n)$$

$$= \sum_{n=0}^{+\infty} P(r(t)\ge x,\ S_n < t \le S_{n+1})$$

$$= \sum_{n=0}^{+\infty} P(S_n<t,\ S_{n+1} \ge t+x) ,$$

where $S_0 = 0$. We now apply the method often utilized in the past, which yields

$$P(S_n<t,\ S_{n+1}\ge t+x) = \int_0^t P(S_{n+1}\ge t+x|S_n = y)dF^{(n)}(y)$$

$$= \int_0^t P(X_{n+1}\ge t+x-y)dF^{(n)}(y) = \int_0^t [1-F(t+x-y)]dF^{(n)}(y) ,$$

where, as usual, we put $X_{n+1} = S_{n+1} - S_n$. Substituting this last expression into (35) and interchanging summation and integration, we obtain the required expression for $R_t(x)$. □

Lemma 4.3.3. With the notations of Lemma 4.3.2, the following recursive formula is valid:

$$P(r(t)\ge x) = 1-F(t+x) + \int_0^t P(r(t-y)\ge x)dF(y) .$$

Proof: In the formula

$$(36)\quad P(r(t)\geq x) = \int_0^{+\infty} P(r(t)\geq x|S_1 = y)dF(y) ,$$

we cut the integration at $y = t$. When evaluating this integral for $t \leq y$, we make use of the fact that $r(t) \geq x$ means that $S_1 \geq t+x$. Therefore

$$P(r(t)\geq x|S_1 = y) = \begin{cases} 1 & \text{if } t+x\leq y \\ 0 & \text{if } t\leq y<t+x , \end{cases}$$

which implies

$$(37)\quad \int_t^{+\infty} P(r(t)\geq x|S_1 = y)dF(y) = \int_{t+x}^{+\infty} dF(y) = 1-F(t+x) .$$

On the other hand, if $y < t$, then the condition $S_1 = y$ means that the origin can be shifted to y and thus $r(t)$ becomes $r(t-y)$ in this new coordinate system. Hence, we have

$$\int_0^{+\infty} P(r(t)\geq x|S_1 = y)dF(y) = \int_0^{t} P(r(t-y)\geq x)dF(y) .$$

Substituting this last equation and (37) into (36), we obtain the recursive formula as stated in the lemma. This completes the proof. □

Lemma 4.3.4. With the notations of Lemma 4.3.2, the joint distribution of $r(t)$ and $a(t)$ is given by

$$(38)\quad P(r(t)\geq x,\ a(t)<z) = \int_{t-z}^{t} [1-F(t+x-y)]dU(y) ,$$

where $x \geq 0$ and $0 < z < t$.

Remark. When $z \geq t$, then the event $\{a(t)<z\}$ becomes the sure event and thus

$$\{r(t)\geq x,\ a(t)<z\} = \{r(t)\geq x\} .$$

Notice that, for $z = t$, the formula of Lemma 4.3.4 indeed reduces to that of Lemma 4.3.2.

As a consequence of Lemma 4.3.4, we obtain the distribution of $a(t)$. Namely, substituting $x = 0$ into (38) we obtain

$$(39)\quad P(a(t)<z) = \begin{cases} \int_{t-z}^{t} [1-F(t-y)]dU(y) & \text{if } 0 < z < t \\ 1 & \text{if } z \geq t . \end{cases}$$

Proof of Lemma 4.3.4. We shall first evaluate

$$P(r(t)\geq x,\ a(t)<z,\ N(t) = n | S_n = y)\ ,\quad n \geq 1.$$

This conditional probability is evidently zero if y does not satisfy the inequalities $t-z<y<t$. On the other hand, if y belongs to the interval $(t-z,t)$, then, by shifting the origin to y, the requirements $r(t)\geq x$, $a(t)<z$ and $N(t) = n$ given $S_n = y$, lead to the same probability as the unconditional probability of S_1 exceeding $t+x-y$. Hence, for $t > z$,

$$\begin{aligned} P(r(t)\geq x,\ a(t)<z) &= \sum_{n=1}^{+\infty} P(r(t)\geq x,\ a(t)<z,\ N(t) = n) \\ &= \sum_{n=1}^{+\infty} \int_{t-z}^{t} P(r(t)\geq x,\ a(t)<z,\ N(t) = n | S_n = y)dF^{(n)}(y) \\ &= \sum_{n=1}^{+\infty} \int_{t-z}^{t} [1-F(t+x-y)]dF^{(n)}(y) \\ &= \int_{t-z}^{t} [1-F(t+x-y)]d\{\sum_{n=1}^{+\infty} F^{(n)}(y)\}\ . \end{aligned}$$

which is the formula (38). The proof is completed. □

Let us evaluate the integral on the right hand side of (38) for a Poisson process. Here $F(x) = 1-e^{-bx}$ and, by (34) or by direct calculation, $U(x) = 1+bx$. Thus, for $t > z$,

$$\begin{aligned} P(r(t)\geq x,\ a(t)<z) &= \int_{t-z}^{t} e^{-b(t+x-y)}bdy \\ &= e^{-bx} - e^{-b(x+z)} = e^{-bx}(1-e^{-bx})\ . \end{aligned}$$

This conclusion is restated in the following

Corollary 4.3.1. For a Poisson process, r(t) and a(t) are independent. The distribution of r(t) is the same for all $t > 0$ and coincides with the interval distribution $F(x) = 1-e^{-bx}$. The distribution function $F_t(z)$ of a(t) equals F(z) for $z < t$ and equals one for $z \geq t$.

We now analyze the extent to which these properties characterize the Poisson processes among renewal processes. The following results are combinations of those obtained by K.L. Chung (1972), K.B. Erickson and H. Guess (1973), E. Cinlar and P. Jagers (1973) and P.T. Holmes (1974). We start with a result due to Erickson and Guess.

Theorem 4.3.1. Let $\{S_n,\ n\geq 1,\ F(x)\}$ be a renewal process. If r(t) and a(t) are independent for one value $t = t^* > 0$, then the distribution function of $r(t^*)$

coincides with $F(x)$ (recall that throughout this chapter, $F(x)$ is assumed to be continuous).

Proof: Because of the independence of $r(t^*)$ and $a(t^*)$,

$$(40) \quad P(r(t^*)\geq x \mid a(t^*)<z) = P(r(t^*)\geq x) \ .$$

On the other hand,

$$P(r(t^*)\geq x \mid a(t^*)<z) = \frac{P(r(t^*)\geq x,\ a(t^*)<z)}{P(a(t^*)<z)} \ .$$

We can therefore apply (38) and (39) in (40) and (41) to evaluate the distribution of $r(t^*)$. We get

$$(42) \quad P(r(t^*)\geq x) = \frac{\int_{t-z}^{t} [1-F(t+x-y)]dU(y)}{\int_{t-z}^{t} [1-F(t-y)]dU(y)} \ ,$$

where $0 < z < t$. Since the left hand side does not depend on z, we can simplify the expression for the distribution of $r(t^*)$ by evaluating the limit of the right hand side as z tends to zero. For this purpose, we estimate the numerator and the denominator as follows:

$$\int_{t-z}^{t} [1-F(t+x-y)]dU(y) \geq \int_{t-z}^{t} \{1-F[t+x-(t-z)]\}dU(y)$$

$$= [1-F(x+z)][U(t) - U(t-z)] \ ;$$

$$\int_{t-z}^{t} [1-F(t+x-y)]dU(y) \leq \int_{t-z}^{t} [1-F(t+x-t)]dU(y)$$

$$= [1-F(x)][U(t) - U(t-z)] \ ;$$

$$\int_{t-z}^{t} [1-F(t-y)]dU(y) \geq \int_{t-z}^{t} \{1-F[t-(t-z)]\}dU(y)$$

$$= [1-F(z)][U(t) - U(t-z)] \ ;$$

and, finally,

$$\int_{t-z}^{t} [1-F(t-y)]dU(y) \leq \int_{t-z}^{t} dU(y) = U(t) - U(t-z) \ .$$

With these estimates, we get from (42),

$$1-F(x+z) \le P(r(t^*) \ge x) \le \frac{1-F(x)}{1-F(z)} .$$

Since $F(x)$ is continuous, and $F(0) = 0$, letting $z \to 0$ results in

$$P(r(t^*) \ge x) = 1-F(x) ,$$

which was to be proved. □

Theorem 4.3.1 shows that the independence of $r(t)$ and $a(t)$ implies a distributional property. Therefore, if the fact that the distribution of $r(t)$ does not depend on t characterizes the Poisson process then so does the mentioned independence property. Before formulating this possibility into an explicit statement, however, we state a theorem obtained independently by Cinlar and Jagers (1973) and Holmes (1974).

Theorem 4.3.2. Let $\{S_n, n\ge 1; F(x)\}$ be a renewal process, where $F(x)$ is continuous. If $E[r(t)]$ is finite, then $\{S_n\}$ is a Poisson process if, and only if, $E[r(t)]$ does not depend on t.

Corollary 4.3.2. Each one of the following properties characterizes the Poisson processes among renewal processes.

(i) the distribution of $r(t)$ does not depend on t and

$$\int_0^{+\infty} [1-F(x)]dx < +\infty ;$$

(ii) the random variables $a(t)$ and $r(t)$ are independent and

$$\int_0^{+\infty} [1-F(x)]dx < +\infty .$$

Proof of Corollary 4.3.2: Corollary 4.3.1 implies properties (i) and (ii) for Poisson processes. Conversely, if (i) holds, then the common distribution of $r(t)$ is evidently the interval distribution $F(x)$. Hence, (i) reduces to the conditions of Theorem 4.3.2, and thus (i) is a characteristic property of a Poisson process. Moreover, since, in view of Theorem 4.3.1, (ii) implies (i), property (ii) also characterizes the Poisson processes. □

Proof of Theorem 4.3.2.: Since the random variable $r(t)$ is nonnegative, we have from Lemma 1.2.1 that

$$E[r(t)] = \int_0^{+\infty} P(r(t)\ge x)dx = r, \text{ say} .$$

Therefore, integrating the recursive formula of Lemma 4.3.3 with respect to x, we get

$$r = \int_0^{+\infty} [1-F(t+x)]dx + \int_0^{+\infty} \{ \int_0^t P(r(t-y) \geq x) dF(y) \} dx$$

$$= \int_t^{+\infty} [1-F(y)]dy + \int_0^t [\int_0^{+\infty} P(r(t-y) \geq x) dx] dF(y) \ .$$

However, recognizing that the inner integral in the last term on the right hand side is the expectation of $r(t-y)$, which also equals r, we obtain the equation

$$r = \int_t^{+\infty} [1-F(y)]dy + r \int_0^t dF(y) = \int_t^{+\infty} [1-F(y)]dy + rF(t) \ ,$$

or equivalently

$$(43) \quad r[1-F(t)] = \int_t^{+\infty} [1-F(y)]dy \ , \ t > 0 \ .$$

Since $F(x)$ is assumed to be continuous with $F(0) = 0$, this last equation is exactly our basic property (P2) in Section 1.5, from which it follows that $F(x)$ is exponential and thus $\{S_n\}$ is Poisson. The theorem is established. □

While in Theorem 4.3.2, a property of $r(t)$ is assumed for all t, the following theorem, due to Erickson and Guess (1973), leads to a characterization of the Poisson process, in terms of the independence of $r(t)$ and $a(t)$ for one value of t.

Theorem 4.3.3. Let $\{S_n, n \geq 1; F(x)\}$ be a renewal process, where $F(0) = 0$, $F(x)$ is continuous and $F(x) > 0$ for all $x > 0$. Then the process $\{S_n\}$ is Poisson if, and only if, there is one $t = t^* > 0$ such that $r(t^*)$ and $a(t^*)$ are independent.

Erickson and Guess give two proofs for this theorem and both proofs would fit into the general approach of the present monograph. Since we do not want to include both proofs, we give the outline of one and details of the other.

One proof is based on the idea of what we called earlier the method of limit laws. That is, first generate a sequence $t_1 < t_2 < \ldots$ from the single t^* such that the independence of $r(t_j)$ and $a(t_j)$ is valid for all j. Then develop a limit theorem based on the independence of $r(t)$ and $a(t)$ on a sequence $t \to +\infty$ which limit theorem involves $F(x)$. Then the actual limit will determine $F(x)$. This idea is carried out as follows. The authors first observe that if $r(t^*)$ and $a(t^*)$ are independent then so are $r(t_j)$ and $a(t_j)$ for all $t_j = jt^* > 0$, $j \geq 1$. Therefore, in view of Theorem 4.3.1, the distribution of $r(t_j)$ is $F(x)$ for all $t_j = jt^*$, $j \geq 1$, i.e.,

$$(44) \quad P(r(t^*j) < x) = F(x) \ , \ j = 1,2,\ldots \ .$$

On the other hand, a general theorem (which is an easy consequence of the so-called renewal theorem and which we do not prove here; see Feller (1966), p. 347) states

$$(45) \quad \lim_{t=+\infty} P(r(t)<x) = b \int_0^x [1-F(y)]dy$$

for any renewal process even under milder conditions than those assumed about F(x), where

$$\frac{1}{b} = \int_0^{+\infty} [1-F(y)]dy$$

with the understanding that if the integral on the right hand side in this last formula is infinite then b = 0 and b = +∞ when the integral is zero. It follows from (43) and (44) that

$$(46) \quad F(x) = b \int_0^x [1-F(y)]dy, \quad x > 0 ,$$

from which we see that 0 < b < +∞ under our assumptions on F(x). However, (46) is again (P2) of Section 1.5 just as in the case of (43). Namely,

$$b \int_0^x [1-F(y)] = b[\int_0^{+\infty} \cdots - \int_x^{+\infty} \cdots] = b(\frac{1}{b} - \int_x^{+\infty} [1-F(y)]dy) .$$

Hence, F(x) is exponential and thus $\{S_n\}$ is Poisson as claimed.

Before turning to the more detailed proof of Theorem 4.3.3, we remark that a proof similar to the one above (that is, utilizing the renewal theorem and the limit relation (45)) is also applied by Chung (1972) to establish a characterization of Poisson processes through distributional properties of r(t) on some sets of t. We should also add that three of the papers we referred to (those by Chung, Erickson and Guess, and Cinlar and Jagers) arrive at equation (43) or (46) and conclude that its only solution is the exponential distribution without assuming continuity. We therefore reemphasize here once again that the continuity of F(x) does follow from (43) for all x > 0, but a solution for (43) may have a jump at x = 0 if the continuity of F(x) is not assumed.

We now present the second proof of Theorem 4.3.3, which is also due to Erickson and Guess.

Proof of Theorem 4.3.3.: In view of Corollary 4.3.1, we have only to prove that the conditions of the theorem imply that the process $\{S_n\}$ is Poisson.

Since r(t*) and a(t*) are independent,

$$P(r(t^*) \geq x, a(t^*) < z) = P(r(t^*) \geq x)P(a(t^*) < z),$$

which, on account of Lemma 4.3.2 and formulas (38) and (39), is equivalent to the validity of the equation

$$(47) \quad \int_{t-z}^{t} [1-F(t+x-y)]dU(y) = R_t^*(x) \int_{t-z}^{t} [1-F(t-y)dU(y) ,$$

where $0 < z < t$ and $R_t^*(x) = 1-R_t(x)$ as given in Lemma 4.3.2. We now claim that the above equation can hold only if

$$(48) \quad 1-F(t+x-y) = R_t(x)[1-F(t-y)] ,$$

where $x > 0$ and $t-z \le y \le t$, except perhaps a set of y's, the U-measure of which is zero. Namely, if (48) fails for some x,t and y, then, since F is assumed to be continuous, it fails on a whole interval of y; if the left hand side was originally larger than the right hand one, then it remains larger on an interval of y and vice versa. In other words, the failure of (48) for one triplet (t,x,y) implies that there is an interval (a,b), $a < b$, such that (48) fails for all (t,x,y^*), where $a \le y^* \le b$. Assume, for example, that the left hand side is larger. Then

$$(49) \quad 1-F(t+x-y^*) > R_t(x)[1-F(t-y^*)]$$

for all $y^* \in I = [a,b]$. However, U(I) is positive for any interval in view of our assumptions that $F(0) = 0$ and $F(x) > 0$ for all $x > 0$. Therefore, (49) contradicts (47) if we choose a,b and z appropriately. In other words, (48) holds for all $x > 0$ and all $t-z \le y \le t$ (the continuity assumption leads to all y rather than to almost all y with respect to U). If we change the variable y in (48) to $t-y = s$, we get

$$1-F(x+s) = R_t(x)[1-F(s)] .$$

Setting $s = 0$ yields $R_t(x) = 1-F(x)$. Hence, we arrived at the lack of memory equation

$$1-F(x+s) = [1-F(x)][1-F(s)] ,$$

where $x \ge 0$ and $0 \le s \le z$ with some $z > 0$. The basic assumptions on F(x) now imply that the only solution of this equation is the exponential distribution, which was to be proved. □

The conclusions of Theorems 4.3.3 and 4.3.1 and Corollary 4.3.1 can be combined in several different ways. For example, if for a renewal process $\{S_n, n \ge 1; F(x)\}$, r(t) and a(t) are independent for one value of $t > 0$, then each of the following statements holds: (i) r(t) and a(t) are independent for all $t \ge 0$; (ii) the distribution function of r(t) is F(x) for all $t \ge 0$; (iii) $E(r(t)) = E(S_1)$ for all $t \ge 0$ and (iv) the distribution function of a(t) equals F(x) for all $x < t$ (recall our basic assumptions on $F(x)$, which are permanent in this chapter). However, if we replace "all $t \geq 0$" in (ii) - (iv) by "one $t > 0$," it is not known whether the conclusion that $F(x)$ is exponential is still valid. It would, for example, be interesting to prove that either there is no other process besides the Poisson process for which $E[r(t)] = E(S_1)$ (assumed finite), or to give an example for a renewal process when $E[r(t)] = E(S_1)$ for some $t > 0$ but $E[r(t)] \neq E(S_1)$ for some other values of t.

We have not yet discussed characterizations based on $a(t)$ alone. However, with completely similar arguments as for $r(t)$, we can conclude that an appropriate specification of the distribution of $a(t)$ or of $E[a(t)]$ characterizes the Poisson process among renewal processes. Therefore, we state without proof the corresponding theorems on $a(t)$.

Theorem 4.3.4. Let $\{S_n, n\geq 1; F(x)\}$ be a renewal process with continuous interval distribution function $F(x)$. Then, if $r(t)$ and $a(t)$ are independent for a fixed $t > 0$, the distribution of $a(t)$ is $F(x)$ for $x < t$.

Theorem 4.3.5. With the notations of the preceding theorem, define

$$F(x;t) = \begin{cases} F(x) & \text{if } 0 \leq x < t \\ 1 & \text{if } x > t \,. \end{cases}$$

Then the process $\{S_n\}$ is Poisson if, and only if,

$$E[a(t)] = \int_0^t x\,dF(x;t) \;, \quad \text{for all } t > 0 \,.$$

For a further extension, when S_1 may have a distribution different from the intervals $S_j - S_{j-1} \geq 0$, $j \geq 1$, and S_1 may even be negative, see F.L. Garagorry and M. Ahsanullah (1977).

Cinlar and Jagers (1973) deduce from Theorem 4.3.5 (for which they also give a detailed proof) the following result.

Theorem 4.3.6. Let $\{S_n, n\geq 1; F(x)\}$ be a renewal process, where $F(x)$ is continuous. Then this process is Poisson if, and only if,

$$(50) \quad E[a(t)\,|\,N(t) = n) = E(S_1\,|\,N(t) = n)$$

for all $t > 0$ and all $n \geq 1$.

We separate the "only if" part of the theorem and we actually prove the following result.

Theorem 4.3.7. For a Poisson process, the conditional distribution of S_1 and S_n, given $N(t) = n$, $n \geq 1$, is the same as the distribution of the minimum and maximum, respectively, of n independent random variables uniformly distributed on the interval $(0,t)$. That is,

$$P(S_1 < x | N(t) = n) = 1 - (1 - \frac{x}{t})^n ,$$

and

$$P(S_n < y | N(t) = n) = (\frac{y}{t})^n ,$$

where $n \geq 1$ and $0 < x,y < t$.

Remark. A consequence of Theorem 4.3.7 is that, for a Poisson process, the conditional distributions of S_1 and $a(t)$, given $N(t) = n \geq 1$, coincide. Hence, one part of Theorem 4.3.6 follows from Theorem 4.3.7.

Proof of Theorem 4.3.7.: Let $0 < x < t$. By definition

$$(51)\quad P(S_1 \geq x | N(t) = n) = \frac{P(S_1 \geq x,\ N(t) = n)}{P(N(t) = n)} .$$

However,

$$\{S_1 \geq x,\ N(t) = n\} = \{N(x) = 0,\ N(A) = n\} ,$$

where A is the interval (x,t). By Theorem 4.2.3, $N(x)$ and $N(A)$ are independent and the distribution of $N(A)$ is the same as that of $N(t-x)$. Hence, on account of (1) and (51)

$$P(S_1 \geq x | N(t) = n) = \frac{e^{-bx}[b(t-x)]^n e^{-b(t-x)}}{(bt)^n e^{-bt}}$$

$$= (\frac{t-x}{t})^n = (1 - \frac{x}{t})^n ,$$

which proves our claim on S_1.

Arguing similarly, we get

$$P(S_n < y | N(t) = n) = \frac{P(N(x) = n)P(N(t-x) = 0)}{P(N(t) = n)} ,$$

which indeed becomes $(x/t)^n$ after the substitution of (1) into the distribution of $N(x)$, $N(t-x)$ and $N(t)$. The proof is completed. □

We add, without going into detail, that, given $N(t) = n$, $n \geq 1$, the joint distribution of the points $S_1, S_2, \ldots, S_n$ in a Poisson process coincides with the joint distribution of the n order statistics of n independent random variables which are uniformly distributed on the interval $(0,t)$. Its proof is a matter of easy calculation similar to those presented in the case of Theorem 4.3.7.

Proof of Theorem 4.3.6.: In view of Theorem 4.3.7 (see the Remark following the statement of Theorem 4.3.7), we have to prove that if (50) holds for all $t > 0$ and all $n \geq 1$, then the process $\{S_n\}$ is Poisson.

Let, for $n \geq 0$,

$$a_n(t) = \begin{cases} a(t) & \text{if } N(t) = n \\ 0 & \text{otherwise}. \end{cases}$$

Then

$$E[a(t)] = E[a_0(t)] + \sum_{n=1}^{+\infty} E[a_n(t)]$$

$$= t[1-F(t)] + \sum_{n=1}^{+\infty} E[a_n(t)] ,$$

where in the last term on the right hand side we can utilize (50). We get

$$(51) \quad E[a(t)] = t[1-F(t)] + \sum_{n=1}^{+\infty} E(S_{1,n}) ,$$

where

$$S_{1,n} = \begin{cases} S_1 & \text{if } N(t) = n \\ 0 & \text{otherwise}, \end{cases}$$

$n = 1,2,\ldots$. However,

$$(52) \quad \sum_{n=1}^{+\infty} E(S_{1,n}) = E\left(\sum_{n=1}^{+\infty} S_{1,n}\right) = \int_0^t x dF(x) ,$$

because $\sum_{n=1}^{+\infty} S_{1,n} = S_1$ if $N(t) \geq 1$ and the previous sum is zero if $N(t) = 0$. That is, the sum of $S_{1,n}$ is S_1 whenever $S_1 < t$. Substituting (52) into (51), we obtain that

$$E[a(t)] = \int_0^t x dF(x;t) \ , \ t > 0 ,$$

where $F(x;t)$ was introduced in Theorem 4.3.5. Theorem 4.3.5 implies that the process is Poisson. □

It is instructive to compare the meaning of Theorem 4.3.6 with results in Section 3.4. In particular, Lemma 3.4.1 and Theorem 3.4.1 imply that the sequence $E_{n:n}$ uniquely determines the population distribution. Now, Theorem 4.3.7 indicates that $E[a(t)|N(t) = n]$ is related to $E_{n:n}$ in the sense that if the process in question is Poisson then it is exactly $t - E_{n:n}$, where the population distribution is uniform on the interval $(0,t)$. However, for non-Poissonian processes, the notation $E[a(t)|N(t) = n] = t - E_{n:n}$ is not justified because the distribution of S_n, given $N(t) = n$, is not that of the maximum of independent and identically distributed random variables. But if we redefine $E_{n:n}$ as the expected value of the maximum of a set $Y_1, Y_2, \ldots, Y_n$ of (not necessarily independent and/or identically distributed) random variables, then, of course, the expected value of $a(t)$, given $N(t) = n$, is $t - E_{n:n}$. Namely, given $N(t) = n$, select one of the points $S_1, S_2, \ldots, S_n$ at random and call it Y_1. Then select a second point and call it Y_2, etc. Evidently, S_n is the maximum of the Y's. Therefore, Theorem 4.3.6 actually asserts that $E_{n:n}$ characterizes the structure of the sequence Y_j as well as the interval distribution $F(x)$ under the assumption (50). This leads to the following general question: under what general dependence structure does the sequence $E_{n:n}$, $n \geq 1$, characterize a given model? An answer to this question would be interesting in two general areas. One could be within the theory of renewal processes (when (50) is appropriately replaced to lead to a characterization of a process different from the Poisson processes). Another area is to assume that the multivariate distribution of the sequence $Y_1, Y_2, \ldots$ belongs to a given family. As an indication in regard to the types of results that can be expected as well as to the difficulty of such an attempt, see Section 5.3, in particular Theorems 5.3.2 and 5.3.3, which are stated in the form of minima rather than maxima.

We do not discuss further the pre-1973 results on the Poisson process; the reader is referred to the comprehensive discussions by D.J. Daley and D. Vere-Jones (1972) and M. Westcott (1973).

The following type of extension of part (i) of Corollary 4.3.2 was initiated by S. Kotz and N.L. Johnson (1974). Let $\{S_n, n \geq 1; F(x)\}$ be a renewal process and define $r(t)$ as before. However, let us now choose the point $t > 0$ with some specific probability distribution $K(u)$. Our interest is again to relate the distribution of $r(t)$ to $F(x)$ and to seek conditions under which a characterization of the Poisson process is obtained. Kotz and Johnson show that if t is selected conditionally uniformly then the distribution of $r(t)$ is still $F(x)$ for the Poisson process and this property does characterize the Poisson process among all renewal processes. However, the uniformity assumption is not necessary here. V. Isham, D.N. Shanbhag and M. Westcott (1975) obtained the following general result.

<u>Theorem 4.3.8</u>. Let $\{S_n, n \geq 1; F(x)\}$ be a renewal process with $E(S_1) < +\infty$. We assume that $F(0) = 0$ and that $F(x)$ is continuous. Let t be a nonnegative random variable with distribution function $K(u) = P(t<u)$, where $K(0^+) < 1$, which is independent of the process $\{S_n\}$. If the distribution function $V(y) = P(r(t) < y)$ equals $F(y)$ for one fixed $K(u)$ then the process is Poisson.

Conversely, for a Poisson process, $V(y) = F(y)$ for an arbitrary distribution $K(u)$.

Remark. In the original statement, the continuity of $F(x)$ is not required. We included it here only for the sake of conformity with our previous assumptions.

Outline of Proof: Since $V(y) = F(y)$, the random variables $N(t+x) - N(t)$ and $N(x)$ are identically distributed. Consequently, their expectations are also equal and thus, on account of (34),

$$(53) \quad U(x) - 1 = E[N(t+x) - N(t)] = E[U(t+x) - U(t)] ,$$

where, in the last step, we interchanged taking expectations with respect to $\{S_n\}$ and t, respectively. Now, by the above equation and by the method of limit laws, we can easily deduce that $U(x) - 1 = bx$ with some constant $b > 0$. Namely, from (53),

$$E[U(t+x+y) - U(t+y)] = E[U(t+x+y) - U(t)]$$

$$- E[U(t+y) - U(t)] = U(x+y) - U(y) .$$

Hence, if t_1 and t_2 are independent random variables with the same distribution as t then the substitution $y = t_2$ and $t = t_1$ above and (53) yield

$$E[U(t_1+t_2+x) - U(t_1+t_2)] = E[U(x+t_2) - U(t_2)] = U(x) - 1 .$$

Similarly, we obtain by induction that if $t_1,t_2,\dots,t_n$ are independent copies of t and if $T_n = t_1+t_2+\dots+t_n$, then

$$(54) \quad E[U(T_n+x) - U(T_n)] = U(x) - 1$$

for all $n \geq 1$ and all $x > 0$. It therefore requires a limit theorem for the left hand side to determine $U(x)$. Basically the renewal theorem, which we have referred to earlier in connection with (45) (Feller (1966), p. 347), guarantees that the left hand side of (54) converges to x/E, where $E = E(S_1)$. Hence, by (54), $U(x) = 1+x/E$. The inverse Fourier transform of $1+x/E$, in view of (33), yields that $F(x)$ is exponential and thus the process $\{S_n\}$ is Poisson.

The converse part of the statement is a matter of easy calculation, hence its details are omitted. □

4.4. Rarefactions, geometric compounding and damage models.

As the title of the present section suggests, we shall relate three different areas of research which were developed independently of each other in the literature.

Although our aim is to emphasize characterizations in connection with these models, we also want to emphasize that the corresponding theories in other aspects as well are strongly related and, in most cases, coincide.

We first describe the three models mentioned in the title.

Rarefactions of renewal processes. Let $\{S_n, n\geq 1; F(x)\}$ be a renewal process. Suppose each point S_n is erased with probability $1-q$. Expand the time scale by a factor $1/q$. We then obtain a new point process $\{S_n^*, n\geq 1\}$. The above procedure will be called a rarefaction, where the erasure of the points S_n will be assumed to be independent of the process $\{S_n\}$ and also the decisions on erasing distinct points S_n will be made independently of each other. We shall use the notation R_q for a rarefaction procedure; hence $R_q\{S_n\} = \{S_n^*\}$. Evidently,the only case of interest is $0 < q < 1$. Since we shall show that $R_q\{S_n\}$ also is a renewal process, we can also speak of the interval distribution $F^*(x)$ of $\{S_n^*\}$; we also use the notation $R_qF(x) = F^*(x)$ to reflect this fact.

Geometric compounding. Let $T_1,T_2,\ldots$ be independent and identically distributed random variables with common distribution function $F(x)$. Let $m(p)$ be a geometric variable with parameter $0 < p < 1$, that is, $P(m(p) = k) = p^{k-1}(1-p)$, $k = 1,2,\ldots$. We assume that $m(p)$ is independent of the T's. Set

$$qS^* = T_1+T_2+\ldots+T_{m(p)} \ , \quad q = 1-p \ ,$$

and $F^*(x)$ for the distribution function of S^*. We are interested in the relation of $F(x)$ and $F^*(x)$, and, in particular, in the theory of limiting distribution of S^* as $p\to 1$. The fact that we used identical notations to those in the first model is not accidental: the meaning of the concepts will become identical.

Damage models. In the description of the model, we shall use one of the original examples from zoology, which induced the early investigations by C.R. Rao and H. Rubin (1964). Consider the time epochs $0 \leq S_1 < S_2<\ldots$ when a bird lays eggs. A squirrel runs around and, at each time, with the same probability, steals an egg from the nest. Therefore, if the ranger did not see the actual epochs when the eggs were laid nor when some eggs were stolen, then the ranger observes "a damaged number" of eggs at any given time. The interest is to compare the distribution of the number of eggs actually laid with the distribution of the number of eggs in the nest,with the aim to characterize these distributions under simple assumptions.

The strong interrelation among the three models described above is evident. In the damage model, we have a point process $\{S_n\}$, and the eggs remaining are a rarefied set of the S_n. The difference is that we did not assume that the S_n form a renewal process. The relation of the geometric compounding model to the rarefaction model is easily seen if we look at S_1^* in the rarefied sequence. It is, apart from the change of time scale, the sum of a random number of intervals

$S_j - S_{j-1}$, namely, $S_1^* = S_k q$, where k is the first index which was not erased. It is straightforward to compute the distribution of k which turns out to be geometric. Therefore, a result in one model can be translated into one in another model. What is surprising in the development of these three theories is that the initial questions and their solutions, although posed without reference to each other, were also related. We shall return to this after some basic results on the rarefaction model.

The rarefaction model was first introduced in the literature by A. Rènyi (1956). All basic results which follow are due to him.

Theorem 4.4.1. For a renewal process $\{S_n, n\geq 1; F(x)\}$, the rarefied process $R_q\{S_n\} = \{S_n^*\}$ is also a renewal process. Its interval distribution function $F^*(x)$ can be computed as follows

$$(55)\quad F^*(x) = q\sum_{k=1}^{+\infty} F^{(k)}(\frac{x}{q})p^{k-1}$$

where $0 < q < 1$, $p = 1-q$ and $F^{(k)}$ is the k-fold convolution of F. Furthermore, $E(S_1) = E(S_1^*)$.

Proof: Let us first evaluate the distribution of S_1^*. From the start, we use the notation $F^*(x)$ for this distribution function. By the total probability rule

$$F^*(x) = \sum_{k=1}^{+\infty} P(S_1^*<x|S_1^* = qS_k)P(S_1^* = qS_k) .$$

Now, the condition $S_1^* = qS_k$ means that $S_1,S_2,\dots,S_{k-1}$ were erased and S_k is preserved. Since the deletions are independent and independent of the original process, the last formula becomes

$$F^*(x) = \sum_{k=1}^{+\infty} P(S_k < \frac{x}{q})p^{k-1}q = q\sum_{k=1}^{+\infty} F^{(k)}(\frac{x}{q})p^{k-1}$$

as claimed.

Let us now calculate the joint distribution of S_1^* and $S_2^* - S_1^*$. Applying the total probability rule again, we get

$$P(S_1^*<x,\ S_2^*-S_1^*<y) = \sum_{k,m} P(S_1^*<x,\ S_2^*-S_1^*<y|B_{k,m})P(B_{k,m}) ,$$

where

$$B_{k,m} = \{S_1^* = qS_k,\ S_2^* = qS_m\},\quad 1 \leq k < m .$$

Since

$$P(B_{k,m}) = p^{k-1}qp^{m-k-1}q = p^{m-2}q^2 ,$$

the independence assumptions reduce the above formula to

$$P(S_1^*<x,\ S_2^*-S_1^*<y) = \sum_{k=1}^{+\infty}\ \sum_{m=k+1}^{+\infty} P(S_k < \tfrac{x}{q},\ S_m-S_k < \tfrac{y}{q})p^{m-2}q^2$$

$$= q^2 \sum_{k=1}^{+\infty}\ \sum_{m=k+1}^{+\infty} F^{(k)}(\tfrac{x}{q})F^{(m-k)}(\tfrac{y}{q})p^{m-2}$$

$$= q^2 \sum_{k=1}^{+\infty} F^{(k)}(\tfrac{x}{q})p^{k-1} \sum_{m=k+1}^{+\infty} F^{(m-k)}(\tfrac{y}{q})p^{m-k-1}$$

$$= F^*(x)F^*(y)\ .$$

Continuing this argument, induction over n yields that $S_1^*, S_2^*-S_1^*, \ldots, S_n^*-S_{n-1}^*$ are independent with common distribution function F^*. That is, the rarefied process $R_q\{S_n\}$ is a renewal process, whose interval distribution function has the form (55).

Moreover, if $E(S_1) < +\infty$, then

$$E(S_1^*) = \sum_{k=1}^{+\infty} E(S_1^* | S_1^* = qS_k)P(S_1^* = qS_k)$$

$$= q\sum_{k=1}^{+\infty} E(S_k)p^{k-1}q = E(S_1)q^2 \sum_{k=1}^{+\infty} kp^{k-1}\ ,$$

which indeed equals $E(S_1)$. On the other hand, since $E(S_1^*) \geq qE(S_1)$, $E(S_1^*) = +\infty$ if $E(S_1) = +\infty$. This completes the proof. □

Corollary 4.4.1. If $\{S_n,\ n\geq 1;\ F(x)\}$ is a Poisson process then so is $R_q\{S_n\}$ with $R_qF(x) = F(x)$.

Proof: If $F(x) = 1 - e^{-bx}$, then

$$F^{(k)}(x) = \int_0^x \frac{b^k y^{k-1} e^{-by}}{(k-1)!}\, dy\ ,\ x \geq 0,\ k \geq 1.$$

Hence, substituting into the formula (55), we have

$$F^*(x) = q\int_0^{x/q} \left(\sum_{k=1}^{+\infty} \frac{b^k y^{k-1} e^{-by}}{(k-1)!}\, p^{k-1} \right) dy$$

$$= qb \int_0^{x/q} e^{-by} \left(\sum_{k=1}^{+\infty} \frac{(byp)^{k-1}}{(k-1)!} \right) dy$$

$$= 1 - e^{-bx} = F(x)\ .$$

Our claim now follows from Theorem 4.4.1. □

Let us compute the characteristic function $\phi^*(s)$ of $F^*(x)$ from formula (55). If $\phi(s)$ denotes the characteristic function of $F(x)$, then

$$(56)\quad \phi^*(s) = q\sum_{k=1}^{+\infty} \phi^k(qs)p^{k-1} = \frac{q\phi(qs)}{1-p\phi(qs)},$$

which we also write in the form

$$(57)\quad \frac{1}{\phi^*(s)} - 1 = \frac{1}{q}\left(\frac{1}{\phi(qs)} - 1\right).$$

When we want to emphasize the relation of $\phi^*(s)$ to $\phi(s)$, we use the notation $R_q\phi = \phi^*$, just as we did for the interval distributions.

We now show that the relation $R_qF(x) = F(x)$ characterizes the Poisson processes.

<u>Theorem 4.4.2</u>. If, for a renewal process $\{S_n, n\geq 1; F(x)\}$ such that $E(S_1) < +\infty$, $R_qF(x) = F(x)$ for one $0 < q < 1$, then the process is Poisson.

<u>Proof</u>: We use the method of limit laws. First, since $R_qF(x) = F(x)$, we have $R_q\phi = \phi$. Next, we show that

$$(58a)\quad R_q(R_qF(x)) = R_{q^2}F(x),$$

and, in general,

$$(58)\quad R_{q_1}(R_{q_2}F(x)) = R_{q_1q_2}F(x).$$

In other words, successive rarefactions can be replaced by a single rarefaction procedure, its parameter being obtained as q_1q_2. As a matter of fact, a repeated application of (57) immediately yields (58), from which we get (58a) by choosing $q_1 = q_2$. Now, since $R_qF(x) = F(x)$, (58a) implies

$$R_{q^2}F(x) = R_q(R_qF(x)) = R_qF(x) = F(x),$$

and, by induction over n, we get from (58) that

$$R_{q(n)}F(x) = F(x),\quad q(n) = q^n,\ n \geq 1.$$

Hence, from (56)

$$(59)\quad \phi(s) = \frac{q^n\phi(q^n s)}{1-(1-q^n)\phi(q^n s)}, \quad n \geq 1 .$$

Letting $n \to +\infty$, $q^n \to 0$, and thus (59) yields (we write $u = q^n$)

$$\phi(s) = \lim_{u=0} \frac{u\phi(us)}{1-\phi(us) + u\phi(us)} = \lim_{u=0} \frac{\phi(us)}{\frac{1-\phi(us)}{u} + \phi(us)}$$

$$= \frac{\phi(0)}{-\phi'(0)s + \phi(0)} = \frac{1}{1-isB},$$

where $B = E(S_1)$. This is the familiar form of the characteristic function of the exponential distribution function $F(x) = 1-e^{-bx}$, $x \geq 0$ and $b = 1/B$. The theorem is established. □

Let us turn to the geometric compounding model. We recognize in (55) that the distribution of S^* is exactly $F^*(x)$ of (55) and we thus have the following consequence of Theorem 4.4.2.

Corollary 4.4.2. If, in a geometric compounding model, the distribution function of S^* coincides with the distribution of T_1 for one value $0 < p < 1$, then this common distribution is exponential.

The relation between the rarefaction model and the model of geometric compounding was first discussed by I.N. Kovalenko (1965) and further analyzed by T. Szántai (1971). Direct proofs for Corollary 4.4.2 were given independently by B. Arnold (1973) and Azlarov *et al.* (1972). We do not intend to make the bibliography on these topics complete; the reader will find a complete bibliography and further results on these two models in the paper by Daley and Vere-Jones (1972), in the extensive investigations by J. Mogyorodi (1969-1973), and in the works by D. Szasz (1976), D. Szynal (1976) and Serfozo (1977). Although most of the results in these papers are formulated in terms of limit theorems, they reduce to characterizations by the method of limit laws.

We finally turn to the damage models. Our aim again is limited here because the Proceedings of the International Conference on Characterizations, held in Calgary, Canada, in 1974 (it is published as Vol. 3 in the series of *Statistical Distributions in Scientific Work*, Ed.: G.P. Patil, S. Kotz and J.K. Ord, D. Reidel Publishing Company, Dordrecht-Boston, 1975) devotes several papers to this subject area. We mainly want to point out the relation of some characterization theorems based on damage models to rarefaction models.

We first introduce notations. Let N and M be nonnegative integer valued random variables, where $M \leq N$ (in our earlier example, N is the number of laid eggs and M is the number of unstolen ones). Our assumption on the actions of the squirrel is that, given $N = n$, the distribution of M is binomial. Now, if we view N as the number of eggs laid during the period $(0,t)$, say, and $S_1<S_2<\ldots<S_N$ are the actual times when the eggs are laid, then the above distributional assumption on M corresponds to the fact that each S_j has the same probability of being "erased" and the erasures occur independently of each other. The question is now that what additional assumption is required to guarantee that N be Poisson? Within the family of distributions for N which can be obtained as the distribution of $N(t)$ in a renewal process, Theorem 4.4.2 provides an answer to this question. However, it turns out that a simple condition characterizes the Poisson distribution among all distributions for N. The following result is due to C.R. Rao and H. Rubin (1964).

Theorem 4.4.3. Let $M \le N$ be nonnegative integer valued random variables. Assume that, given $N = n$, the distribution of M is binomial with parameters n and $0 < p < 1$. Then N is Poisson if, and only if,

$$(60) \quad P(M = m) = P(M = m|M = N) .$$

Remark. From the description preceding the statement of the theorem it is clear that a characterization of a Poisson distribution for N implies a characterization of the Poisson process by a rarefaction procedure. Therefore, the most general model is a damage model out of the three models of the present section. The possibility of characterizations of Poisson processes by rarefaction, through utilizing known results for damage models was pointed out by R.C. Srivastava (1971), but this relation was not exploited in any other papers.

We do not prove Theorem 4.4.3. Instead, the reader is referred to the paper by G.P. Patil and M.V. Ratnaparkhi (1975), who follow the simplest known method of proof, due to D.N. Shanbhag (1974). Patil and Ratnaparkhi also give several extensions of Theorem 4.4.3, and they give a good collection of references. See also the paper by R.C. Srivastava and J. Singh (1975) in the same Proceedings. V.I. Oliker and J. Singh (1977) make some advance on solving the equation (1.2) of the paper by Srivastava and Singh (1975), but a complete solution is still lacking.

Other significant contributions to characterizations by damage models, besides those in the *Calgary Proceedings* mentioned earlier, are those by S. Talwalker (1970), R.C. Srivastava and A.B.L. Srivastava (1970), D.N. Shanbhag (1977) and D.N. Shanbhag and J. Pararetos (1977). Talwalker gives a bivariate variant of the Rao-Rubin theorem (Theorem 4.4.3), while Srivastava and Srivastava obtain a kind of converse to Theorem 4.4.3. They show that if N is a Poisson variable with parameter λ and if the conditional distribution of M, given $N = n$, does not depend on λ, then this latter (conditional) distribution is binomial if, and only if, the Rao-Rubin condition (60) holds. Shanbhag unifies the results of Rao and Rubin, Talwalker and Srivastava and Srivastava. Implicitly, through his method of proof, Shanbhag also relates renewal theory and damage models. Shanbhag and Panaretos carry the method of Shanbhag (1977) further and show that if the parameter p in Theorem 4.4.3 is allowed to depend on n then its conclusion does not remain valid. They also show that the major assumptions of Srivastava and Srivastava cannot be relaxed. Two simple examples of Shanbhag and Parametos disprove two conjectures raised at the Calgary Conference on Characterizations.

CHAPTER 5

CHARACTERIZATIONS OF MULTIVARIATE EXPONENTIAL DISTRIBUTIONS

5.1. Introduction

The development of multivariate exponential distributions got underway very late mainly due to the following two factors: a) the rather tardy development of univariate exponential models as indicated in Chapter 1, and b) the overwhelming predominance of "the" multivariate normal distribution in both statistical practice and theory. The universal belief in correlation and regression techniques, coupled with the unjustified popularity of univariate normal distributions were responsible for the unchallenged position of "the" multivariate distribution for over 80 years since this distribution involves univariate normal marginals and utilizes the correlation coefficient as the dependence mechanism in its structure.

It is interesting to note that bivariate (and multivariate) extensions related to exponential distributions or their transforms were first developed for extreme-value distributions rather than for the simpler exponential distributions. These extensions appeared in J. Geffroy's lengthy papers (1958, 1959) and in J. Tiago de Oliveira's somewhat shorter contribution (1958) as well as in Sibuya's work in early 1960. These were all motivated by Gumbel's famous treatise (1953, 1954, 1958) on extreme-value distributions in which the rudiments of bivariate and multivariate extensions of extreme value distributions were presented as well. Unfortunately, the above mentioned contributions--which appeard in European and Japanese journals--went unnoticed for some time. In fact, the characterization theorems for multivariate extreme value distributions constitute one of the most complete and developed areas of multivariate characterizations. These are discussed in detail in Chapter 5 of J. Galambos' (1978) book and are therefore not presented in this monograph.

It was only in 1960 that an authoritative paper, specifically devoted to exponential multivariate distributions was published in a journal with wide circulation. E. Gumbel (1960) introduced a number of bivariate exponential distributions mainly as a warning against overreliance on multivariate normal techniques, by indicating that many properties of bivariate exponential distributions are quite different from those of the bivariate normal.

Gumbel's contribution was followed by Freund (1961), but it was the paper by Marshall and Olkin (1967a) that initiated the intensive investigations into the realm of multivariate exponential distributions that we are witnessing today. Since our subject matter is essentially in its infancy it would seem inappropriate to attempt to unify the results obtained. We shall therefore confine ourselves to a critical summary of the characterization theorems available thus far. Our purpose here is somewhat different from our aims in preceding chapters. We would like to stimulate additional work by focusing attention on the difficulties and ambiguities involved in extending univariate concepts into the multivariate domain.

Numerous devices and algorithms for generating multivariate distributions have been developed over the past few decades. Some of these are briefly sketched below without commenting on their relative merits.

1) Chronologically, the earliest method was to formally generalize a *system* or *equation* defining a univariate distribution to the multivariate case. The numerous approaches to extending Pearson's system of distributions, e.g. van Uven (1947), Sagrista (1962), Steyn (1960), can be cited as examples of this approach.

2) Another method is to define a multivariate distribution by explicitly specifying the mathematical relations between the joint distribution and its marginals. This was initiated by Fréchet (1951), and more recently extended by Morgenstern (1956), Gumbel (1960), Farlie (1960) and Johnson and Kotz (1975a) among others.

3) Still another approach is to postulate a specified multivariate form for the density by reproducing in the multivariate setting the functional form of a particular univariate family of densities. This includes linear-exponential type distributions originated by Bildikar and Patil (1968), quadratic-exponential type by Day (1968) and multivariate θ-generalized distributions of Goodman and Kotz (1973).

4) The most direct approach--*modelling*--is presented in the works of Freund (1961) and Marshall and Olkin (1967a, 1967b), whereby a multivariate distribution is obtained by indicating a specific stochastic model comprising a system of several "components" representing the random variables under consideration. The multinomial distribution is a primary "generic" example of such a construction.

5) Finally an even sounder aqproach is to extend (hopefully uniquely) a meaningful characterizing property of the univariate distribution to the multivariate case. It is therefore the authors' strong belief that characterization theorems should play a decisive role in determining the multivariate extensions of univariate distributions. In this connection we should be mindful of the fact that, in general, there are infinitely many multivariate distributions which have specified one-dimensional marginal distributions; this observation is especially valid in both the normal and (in spite of some misconceptions to the contrary) exponential cases.

5.2. Bivariate extensions based on LMP

In previous chapters we have shown the equivalence between various characterizing properties of the univariate exponential distribution and its monotone transforms. These equivalent properties will serve as the departure points for building appropriate multivariate distributions and at the same time may provide a characterizing theorem. As we shall see, the equivalent properties are generalized and extended in different directions and result in distinct classes of multivariate exponential distributions.

The first basic characterizing property (P3) of the univariate exponential distribution, that of the *Lack of Memory Property* (LMP) was extended in the pioneering paper of Marshall and Olkin (1967a) and that of Basu and Block (1974).

The univariate statement (in the notation of the previous chapters)

(1) $G(s+t) = G(s)G(t)$

for all $s \geq 0$ and $t \geq 0$ can be extended in various ways. One obvious extension is

(2) $G(s_1+t_1,s_2+t_2) = G(s_1,s_2)G(t_1,t_2)$

for all $s_1 > 0$, $s_2 > 0$, $t_1 > 0$ and $t_2 > 0$.

Unfortunately it is too strong a generalization as the following theorem indicates:

Theorem 5.2.1. The only solution of equation (2) is

$$G(s,t) = \exp\{-(\lambda_1 s+\lambda_2 t)\}$$

for some $\lambda_1 > 0$ and $\lambda_2 > 0$. (In other words, we obtain the "trivial" bivariate exponential distribution which is a product of its marginals, thus corresponding to a device consisting of two independent components.)

Proof: Setting $s_2 = t_2 = 0$ in (2) we immediately obtain

$$G_1(s_1+t_2) \equiv G(s_1+t_1,0) = G_1(s_1)G_1(t_1)$$

for all s_1, $t_1 \geq 0$, which as we know from Chapter 1 implies that $G_1(s) = \exp(-\lambda_1 s)$ for some $\lambda_1 > 0$. Analogously, we have $G_2(t) \equiv G(0,t) = \exp(-\lambda_2 t)$ for some $\lambda_2 > 0$. Now choosing $s_2 = t_1 = 0$, we have

$$G(s_1,t_2) = G_1(s_1)G_2(t_2)$$

as claimed. □

We are therefore guided to relax the requirement for the validity of (2) to a smaller set of values of s_i,t_i, $i = 1,2$.

A physically meaningful relaxation of (2) is the following

(2a) $G(s_1+t,\ s_2+t) = G(s_1,s_2)G(t,t)$

for all $s_1 \geq 0$, $s_2 \geq 0$ and $t \geq 0$.

Requirement (2a) states that if two components in a two-component system reached ages exceeding s_1 and s_2, then the probability that both components are still functioning t time units later is the same as if both were new. In terms of two random variables X and Y (2a) can be written as:

(2b) $P(X>s_1+t, Y>s_2+t \mid X>s_1, Y>s_2) = P(X>t,\ Y>t)$

for all $s_1 \geq 0$, $s_2 \geq 0$ and $t \geq 0$. The following remarkable result is due to Marshall and Olkin (1967a):

Theorem 5.2.2. The unique solution of equation (2b) (or that of (2a)) under the assumption that the marginal distributions of X and Y are exponential is:

$$(3) \quad G(s,t) = P(X>s, Y>t) = \exp\{-\lambda_1 s-\lambda_2 t-\lambda_{12}\max(s,t)\}$$

where λ_1, λ_2 and λ_{12} are nonnegative.

Remark: This is the bivariate exponential distribution of Marshall and Olkin (which can also be derived, as noted above, from several distinct plausible modelling considerations, some of which are described below).

Proof: Setting $s_1 = s_2$ into (2a) yields

$$G(s+t,s+t) = G(s,s)G(t,t)$$

which implies that $G(s,s) = \exp(-\delta s)$ for some $\delta > 0$. Therefore

$$G(s+t,t) = G(s,0)e^{-\delta t} .$$

Now since the marginals of X and Y are assumed to be exponential we have

$$G(s+t,t) = \exp(- (\theta_1 s+\delta t)) \quad \text{for some } \theta_1 \geq 0$$

which means that (since $t \geq 0$)

$$(4a) \quad G(x,y) = \exp(-\delta y-\theta_1(x-y)) \quad \text{for } x \geq y .$$

Analogously we obtain that

$$(4b) \quad G(x,y) = \exp(-\delta x-\theta_2(y-x)) \quad \text{for } x \leq y$$

(for some $\theta_2 \geq 0$).

(4a) and (4b) together define a bivariate distribution for all $x,y \geq 0$. Since $G(x,y)$ is monotonically decreasing in y, we have $\delta \geq \theta_1$, hence $\lambda_2 = \delta-\theta_1$ is nonnegative. Similarly, since $G(x,y)$ is decreasing in x, $\delta \geq \theta_2$, thus $\lambda_1 = \delta-\theta_2$ is also nonnegative. Now let $\lambda_{12} \equiv \theta_1+\theta_2-\delta$. To show that $\lambda_{12} \geq 0$ we must show that $\theta_1+\theta_2 \geq \delta$. However, the bivariate distribution as given by (4a) and (4b) generates the univariate distribution given by the distribution function

$$F(x) = F(x,x) = 1-G(x,x) + G_1(x) + G_2(x)$$

where $G_1(x) = e^{-\theta_1 x}$, $G_2(x) = e^{-\theta_2 x}$ and $G(x,x)$ is given by (4a) and (4b) with $y = x$. Thus

$$f(x) = F'(x) = \theta_1 e^{-\theta_1 x} + \theta_2 e^{-\theta_2 x} - \delta e^{-\delta x} \geq 0$$

for all x. (Here we have essentially utilized the fact that for any two points (x_1,y_1) and (x_2,y_2) a bivariate survival function must satisfy

$G(x_1,y_1) + G(x_2,y_2) - G(x_1,y_2) - G(x_2,y_1) \geq 0$.)

Letting $x \to 0$ we observe that $\theta_1 + \theta_2 - \delta \geq 0$ which shows that λ_{12} is nonnegative. Substituting now $\delta = \lambda_1 + \lambda_2 + \lambda_{12}$, $\theta_1 = \lambda_1 + \lambda_{12}$ and $\theta_2 = \lambda_2 + \lambda_{12}$ into (4a) and (4b) we obtain that the unique solution for equation (2a) is indeed the distribution given by (3), the Marshall-Olkin bivariate exponential distribution. Observe that the nonnegativity of λ_{12} is a consequence of the "rectangular positivity" of any bivariate distribution which is not necessarily a consequence of the monotonicity of $G(\cdot)$ in each one of the variables separately.

The bivariate exponential distribution of Marshall and Olkin is not absolutely continuous, in fact it involves a *singular component* $G_s(x,y)$. Namely,

Theorem 5.2.3. The survival $G(x,y)$ given by (3) can be written as the mixture

$$(5) \quad G(x,y) = \frac{\lambda_1+\lambda_2}{\delta} G_a(x,y) + \frac{\lambda_{12}}{\delta} G_s(x,y)$$

where $G_s(x,y) = \exp[-\lambda \max(x,y)]$ and as above $\delta = \lambda_1 + \lambda_2 + \lambda_{12}$ and the *absolutely continuous part* is given by

$$(6) \quad G_a(x,y) = \frac{\delta}{\lambda_1+\lambda_2} \exp[-\lambda_1 x - \lambda_2 y - \lambda_{12} \max(x,y)] - \frac{\lambda_{12}}{\lambda_1+\lambda_2} \exp[-\delta \max(x,y)] \ .$$

Proof: The decomposition $G(x,y) = \alpha G_a(x,y) + (1-\alpha)G_s(x,y)$ and (3) yields

$$\frac{\partial^2 G(x,y)}{\partial x \partial y} = \alpha f_a(x,y) = \begin{cases} \lambda_2(\lambda_1+\lambda_{12})G(x,y) & \text{if } x > y \\ \lambda_1(\lambda_2+\lambda_{12})G(x,y) & \text{if } x < y \ , \end{cases}$$

where $f_a(x,y)$ is the corresponding two dimensional density and the integration of $\alpha f_a(x,y)$ yields the value of α as given in (5). □

Observe that the event $X = Y$ has, in this case, a positive probability, while the two-dimensional Lebesgue measure of the line $x = y$ is 0. The presence of the singular component is in a sense a "natural phenomenon" as the following theorem shows.

Theorem 5.2.4. The assumption of absolute continuity in addition to assumptions of the lack of memory property and exponential marginals yield a unique bivariate distribution with independent exponential marginals (identical to the one obtained as a solution of equation (1) above).

Proof: Observe that the basic equation (2a) can be rewritten in an equivalent form

$$(2c) \quad G(x,y) = \begin{cases} e^{-\delta y} G_1(x,y) & \text{if } x \geq y \geq 0 \\ e^{-\delta x} G_2(y-x) & \text{if } y \geq x \geq 0 \end{cases}$$

for some $\delta > 0$, where $G_1(x) = G(x,0)$, $G_2(y) = G(0,y)$ and $G(0,0) = 1$. Indeed, setting $(s_1,s_2) = (s,s)$ into (2a) we have

$$G(s+t,s+t) = G(s,s)G(t,t);$$

hence as above we have

$$G(s,s) = \exp(-\delta s) \quad \text{for some } \delta > 0.$$

Now setting $s_2 = 0$ into (2a) we obtain

$$G(s_1+t,t) = G(s_1,0)G(t,t) = G(s_1,0)\exp(-\delta t) .$$

Noting that $t \geq 0$ we have the first part of the statement (2b). The second part follows analogously. □

Moreover, we also have the following result.

Theorem 5.2.5. If the marginals are assumed to have absolutely continuous densities then $G(x,y)$ given by (2b) is absolutely continuous if

(7) $\delta = f_1(0) + f_2(0)$

where $f_j(\cdot)$, $j = 1,2$, are the corresponding densities.

Proof: Recall that

$$G(x,y) = \alpha G_a(x,y) + (1-\alpha)G_s(x,y) \ , \ 0 \leq \alpha \leq 1$$

where $G_a(\cdot)$ and $G_s(\cdot)$ are as above, the absolute continuous and singular components of $G(\cdot)$ respectively. (The singular component is located in the line $x = y$ so it can be obtained from $(1-\alpha)G_s(x,x) = G(x,x) - \alpha G_a(x,x)$.) From (2b), taking the second derivative $\frac{\partial^2 G(x,y)}{\partial x \partial y} = \alpha f_a(x,y)$ we have, in particular,

$$\int_{x \geq y} \alpha f_a(x,y)dx\,dy = 1 - \frac{1}{\delta} f_1(0)$$

$$\int_{x \leq y} \alpha f_a(x,y)dx\,dy = 1 - \frac{1}{\delta} f_2(0) ,$$

and since $F_a(\infty,\infty) = 1$, we obtain

$$\alpha = \int_0^\infty \int_0^\infty \alpha f_a(x,y)dx\,dy = 2 - \frac{1}{\delta} [f_1(0)+f_2(0)] ;$$

hence if $G(\cdot)$ is absolutely continuous and thus $\alpha = 1$, (7) follows. □

Now we are ready to prove the basic assertion due to Block and Basu (1974).

Theorem 5.2.6. If (X,Y) is a positive bivariate random vector which is absolutely continuous and possesses the LMP property, then the exponentionality of the marginals implies that both X and Y are independent.

Proof: Let the marginals be

$$f_1(x) = a_1 \exp(-a_1 x) \quad \text{and}$$

$$f_2 = (y) = a_2 \exp(-a_2 y) \quad \text{for } x,y > 0 ,$$

where a_1 and $a_2 > 0$. From the above we obtain

$$f_1(0) + f_2(0) = a_1 + a_2 = \delta$$

and hence from (2b)

$$G(x,y) = \begin{cases} \exp(-(a_1+a_2)y)\exp(-a_1(x-y)) & \text{if } x \geq y \geq 0 \\ \exp(-(a_1+a_2)x)\exp(-a_2(y-x)) & \text{if } y \geq x \geq 0 . \end{cases}$$

In other words

$$G(x,y) = \exp(-a_1 x)\exp(-a_2 y) \quad \text{for } x,y \geq 0$$

which shows that X and Y are independent. □

To obtain yet another characterization of a bivariate Marshall-Olkin exponential distribution we shall observe that for this distribution as given by (3), we have

$$P(\min(X,Y) > s) = G(s,s) = e^{-\delta s} , \quad s \geq 0 ,$$

where, as above, $\delta = \lambda_1 + \lambda_2 + \lambda_{12}$.

In other words, a *part* of the univariate characterizing property (P4*)--for any two independent exponential random variables the minimum is distributed exponentially--continues to hold for this particular bivariate extension. In fact we have

Theorem 5.2.7. (X_1,X_2) is a bivariate M-0 exponential distribution if and only if there exist exponential random variables U_1, U_2 and U_{12} such that $X_1 = \min(U_1,U_{12})$ and $X_2 = \min(U_2,U_{12})$.

(Note, however, that the property that $\min(X_1,X_2)$ is exponentially distributed does not generally imply that (X_1,X_2) has a bivariate M-0 exponential distribution as the following example indicates:

Let (X_1,X_2) satisfy $P(X_1 \geq x_1, X_2 \geq x_2) = e^{-x_1-x_2+2(1/x_1 + 1/x_2)^{-1}}$, $x_i \geq 0$, $i=1,2$.

In this case $P(\min(X_1,X_2) \geq x) = e^{-x}$, $x \geq 0)$.

Proof: Although a direct proof of this assertion is straightforward, it is instructive to couch it in the language of the *fatal shock model* which serves as a basis for this distribution.

Consider a system consisting of two dependent components. Shocks of three types are present in the environment and arrive independently and at random (i.e. three independent Poisson processes govern the occurrence of shocks). A shock of the first type destroys component 1; it has an expected rate of arrival λ_1, i.e. it arrives at random time U_1 where $P(U_1>t) = e^{-\lambda_1 t}$, $t \geq 0$. Similarly, a shock of type two, with expected rate of arrival λ_2 arriving at random time U_2 where $P(U_2>t) = e^{-\lambda_2 t}$, $t \geq 0$, destroys the second component. Moreover, both components fail if they are subject to a shock of type three with expected rate λ_{12} arriving at random time U_{12}, where $P(U_{12}>t) = e^{-\lambda_{12}t}$, $t \geq 0$. Therefore the (random) life-length T_i of component i = 1,2, satisfies

$$X_i = \min(U_i,U_{12}) \quad i = 1,2,$$

and the joint survival function:

$$G(t_1,t_2) = P(X_i > t_i,\ i=1,2)$$

(8)
$$P[\text{no type 1 arrivals at time } t_1] \times$$
$$P[\text{no type 2 arrivals at time } t_2] \times$$
$$P[\text{no type 3 arrivals at time } \max(t_1,t_2)]$$
$$= e^{-\lambda_1 t_1-\lambda_2 t_2-\lambda_{12}\max(t_1,t_2)}, \quad (\text{for } t_1 \geq 0,\ t_2 \geq 0)$$

which is identical to (3).

From this model it is obvious that if X_1 and X_2 have survival distribution (3) then there exist independent exponential variables U_1,U_2 and U_{12} satisfying $X_1 = \min(U_1,U_{12})$ and $X_2 = \min(U_2,U_{12})$ and conversely. □

(Note: This representation also yields an alternative proof of the decomposition G(x,y) as given by (5) into an absolutely continuous and singular component (cf. eqs. (5) and (6)); the singular component is

$$P[X_1>t_1,\ X_2>t_2 | U_{12} \leq \min(U_1,U_2)] ,$$

and the absolute continuous one is the conditional distribution

$$P[X_1>t_1,\ X_2>t_2 | U_{12} > \min(U_1,U_2)] .$$

Finally, we observe that recently Block (1977) succeeded in providing a new characterization of the M-O bivariate exponential distribution. This characterization is of the type developed by Ferguson (1964) for the univariate exponential distribution (see Chapter 3).

Block's characterization of the M-O bivariate exponential distribution states:

Theorem 5.2.8. A bivariate random variable (X,Y) has the survival function given by (3) iff:

a) (X,Y) has exponential marginals
b) $U \equiv \min(X,Y)$ is exponential
c) $U \equiv \min(X,Y)$ is independent of $V \equiv X-Y$.

Proof: To prove this result it is sufficient to note that an equivalent statement would be: for a nonnegative bivariate random variable (X,Y) with absolutely continuous marginals and U and V as defined above, the LMP holds iff:

(i) U and V are independent
(ii) U is exponential with mean δ^{-1} for some δ.

(Indeed, recall that the M-0 bivariate exponential distribution is the only bivariate distribution with *exponential marginals* satisfying (2a).) The *if* part of the last statement follows from direct computations of

$$P(U = \min(X,Y) \le s,\ V = X-Y \le t)$$

recalling that the LMP implies the existence of $\theta > 0$ such that $P(U = \min(X,Y) > s) = \exp(-\theta s)$, $s > 0$. The *only if* part is, however, more illuminating. Indeed, assuming that U and V are independent and that U is exponential we arrive at:

$$(9)\quad G(x,y) = P(X>x,\ Y>y) = P(U>y) + \int_x^y P(V\le u-y)\,dP(U<u)$$

$$= \exp(-\theta y) + \int_x^y P[V\le u-y]\cdot\theta\exp(-\theta u)\,du\,,\quad \text{for } 0 \le x \le y\,.$$

Consequently, for a fixed $y > 0$ and $0 \le x \le y$ such that $P(V\le x-y)$ is continuous we have

$$\frac{\partial}{\partial x} G(x,y) = -P(V\le x-y)\theta\exp(-\theta x)\,.$$

Equivalently, recalling that $G(t,t) = P(U>t)$,

$$(10)\quad \frac{\partial}{\partial s_1} G(s_1+t,\ s_2+t) = G(t,t)\frac{\partial}{\partial s_1} G(s_1,s_2)$$

is valid for all s_1 (except for at most a countable number of them) such that $0 \le s_1 \le s_2$, $0 \le t$, with some $s_2 > 0$ and $t \ge 0$ fixed. Finally, integrating (10) with respect to s_1 over $0 \le s_1 \le s_2$ we obtain that equation (2a) is valid on the range $0 \le s_1 \le s_2$, $0 \le t$. Similar computations verify the validity of (2a) for $0 \le s_2 \le s_1$ and $t \ge 0$. □

Observe that the assumption that $U \equiv \min(X,Y)$ is exponential is essential for the above characterization. This is because a nonexponential choice for the distribution of U implies that $G(t,t) = P(U>t)$ is nonexponential, which in general contradicts the LMP. Compare also with the example presented following the statement of Theorem 5.2.7.

A bivariate Weibull distribution was introduced briefly by Marshall and Olkin (1967) and was studied in detail by Moeschberger (1974). The distribution is of the form

$$(11)\quad (X_1',\ X_2') = (X_1^{1/\alpha_1},\ X_2^{1/\alpha_2})$$

where $\alpha_i > 0$, $i = 1,2$, and (X_1,X_2) has the M-0 bivariate exponential distribution.

The only characterization theorem for this distribution available so far is the extension of Theorem 5.2.7 which is valid provided $\alpha_i = \alpha$ $(i = 1,2)$.

<u>Theorem 5.2.7a</u>. $(X_1',\ X_2')$ is a bivariate Weibull distribution given by

(12) $G(x_1,x_2) = \exp\{-\lambda_1 x_1^\alpha - \lambda_2 x_2^\alpha - \lambda_{12}\max(x_1^\alpha,x_2^\alpha)\}$, $x_i \geq 0$, $\alpha > 0$,

if and only if there exist independent Weibull random variables U_1', U_2' and U_{12}' such that $X_1' = \min(U_1', U_{12}')$ and $X_2' = \min(U_2', U_{12}')$.

The *proof* of the direct part is a straightforward verification that if U_1', U_2', U_{12}' are independent Weibull random variables with equal shape parameters and $X_1' = \min(U_1',U_{12}')$ and $X_2' = \min(U_2',U_{12}')$, then (X_1',X_2') is bivariate Weibull. The reverse relationship follows from the definition of the distribution via (11) and its validity for the M-O exponential bivariate distribution, or employing a "contrapositive" argument as suggested by Moeschberger (1974).

5.3. The Multivariate Marshall-Olkin exponential distribution and its extensions

The bivariate Marshall-Olkin exponential distribution can be generalized to n dimensions but the notation becomes rather formidable. First, however, we note that the extension of property (2) to n dimensional (the strong lack of memory property) will lead us, in this case, to a multivariate exponential distribution which is a product of its (exponential) marginals, thus forming a system consisting of n independent components, with the survival function of the form

$$G(t_1,t_2,\ldots,t_n) = \exp\{-\sum_{i=1}^{n} \lambda_i t_i\} \text{ , } t_i \geq 0;\ \lambda_i > 0,\quad i = 1,\ldots,n \text{ .}$$

On the other hand,

Theorem 5.3.1. The extension of the "no deterioration" property as given by (2a) coupled with the requirement that the marginals are "exponentials" yields an n-dimensional Marshall-Olkin exponential distribution with the joint survival distribution

$$(13)\quad G(t_1,\ldots,t_n) = \exp[-\sum \lambda_i t_i - \sum_{i<j} \lambda_{ij} \max(t_i,t_j) - \sum_{i<j<k} \lambda_{ijk}\max(t_i,t_j,t_k) - \ldots - \lambda_{12\ldots n}\max(t_1,\ldots,t_n)] \text{ ,}$$

where $t_i \geq 0$ and the λ's are nonnegative constants.

Proof: It is sufficient to show that the n-dimensional distribution given by (13) is the only distribution with (n-1)-dimensional marginals of the same form for which the joint survival distribution of a set of n components each of age t is the same as that of a set of new components. Now this property can be written as:

$$P[X_1>s_1+t,\ldots,X_n>s_n+t \mid X_1>t,\ldots,X_n>t] = P[X_1>s_1,\ldots,X_n>s_n]$$

or

$$G(s_1+t,\ldots,s_n+t) = G(s_1,\ldots,s_n)G(t,\ldots,t);\quad t \geq 0,\ \forall\ s_i \geq 0 \text{ .}$$

As in the bivariate case, setting $s_1=s_2=\ldots=s_n=s$, we have $G(s,\ldots,s) = e^{-\delta s}$ for some $\delta > 0$. Next setting $s_n = 0$ we have

$$G(s_1+t,\dots,s_{n-1}+t,t) = G(s_1,\dots,s_{n-1},0)G(t,\dots,t) = e^{-\lambda t}G_{n-1}(s_1,\dots,s_{n-1}) ,$$

where $G_{n-1}(\cdot)$ is an (n-1)-dimensional marginal and thus, by assumption, is a Marshall-Olkin distribution given by (13) with n replaced by n-1. Using this assumption we obtain that (13) is valid for all $s_i \geq 0$ such that $s_n \leq s_i$, $i = 1,\dots,n-1$ (since $t \geq 0$). However, by symmetry (13) holds for all $s_i \geq 0$, $i = 1,\dots,n$. □

A number of characterizations of the distribution (13) are known, since most of the bivariate characterizations discussed above are easily extendable to the general case. For example,

Theorem 5.3.2. A random vector $(X_1,\dots,X_n)$ has the (n-dimensional) Marshall and Olkin exponential distribution if and only if there exists a collection H_J, $J\epsilon\mathcal{J}$ (where $\mathcal{J}$ is the class of nonempty subsets of $\{1,2,\dots,n\}$) of *independent* exponential random variables such that $X_i = \min(H_J,\ J\epsilon\mathcal{J},\ i\epsilon J)$, $i = 1,\dots,n$.

More explicitly, denote by V the set of vectors $\underset{\sim}{v} = (v_1,\dots,v_n)$ where each v_i is 0 or 1 but $(v_1,\dots,v_n) \neq (0,\dots,0)$. Equation (13) can then be rewritten as

$$\text{(13a)}\quad G(t_1,\dots,t_n) = \exp[-\sum_V \lambda_{v_1,\dots,v_n} \max(t_1v_1,\dots,t_nv_n)],\quad t_i \geq 0 .$$

The characterization of (13a) in terms of minima asserts the existence of 2^n-1 independent exponential random variables $Z_{\underset{\sim}{v}}$, $\underset{\sim}{v}\epsilon V$ such that $X_i = \min_V\{Z_v | \underset{\sim}{v}_i = 1\}$.

Proof: The claim of the theorem is an immediate extension of the corresponding two-dimensional property discussed in the previous section and has especially intuitive verification if we define the multivariate Marshall and Olkin exponential distribution via the fatal shock model. In other words we postulate independent Poisson processes $Z_i(t)$ with rates λ_i which govern the occurrence of fatal shocks to component i, for $i = 1,2,\dots,n$, Poisson processes $Z_{ij}(t)$ with rates λ_{ij} governing the occurrence of fatal shocks to components i and j simultaneously for $1 \leq i \leq j \leq n,\dots,$ and the Poisson process $Z_{12\dots n}(t)$ with the rate $\lambda_{12\dots n}$ which governs the occurrence of fatal shocks to all n components simultaneously. The joint survival distribution for this model is given by (13). Using this model the existence of independent exponential random variables $Z_{\underset{\sim}{v}}$, $\underset{\sim}{v}\epsilon V$ satisfying $X_i = \min_{\underset{\sim}{v}}\{Z_{\underset{\sim}{v}} | v_i = 1\}$ follows in exactly the same manner as in the two-dimensional case. □

Note: Observe also that

$$X_1 = \min\{Z_{\underset{\sim}{v}}\colon v_1=1\} = \min[\min\{Z_{\underset{\sim}{v}}\colon v_1=1,v_2=0\},\ \min\{Z_{\underset{\sim}{v}}\colon v_1=1,v_2=1\}] .$$

This characterizing property has important implications in reliability theory in view of the concept introduced by Esary and Marshall (1974). These authors defined a set of nonnegative random variables $(Y_1,\dots,Y_n)$ to have a joint distribution with *exponential minima* if

$$P[\min_{i\in J} Y_i > t] = e^{-\theta_J t}, \quad t \geq 0$$

for some $\theta_J > 0$ and all nonempty sets $J \subset \{1,\dots,n\}$. Clearly, the Marshall and Olkin distribution is a joint distribution with exponential minima. Moreover, a rather straightforward argument shows that

Theorem 5.3.3. The joint distribution of variables $(Y_1,\dots,Y_n)$ with exponential minima can be marginally equated in minima to a Marshall-Olkin distribution of variables $(X_1,\dots,X_n)$ in the sense that

$$P[\min_{i\in J} Y_i > t] = P[\min_{i\in J} X_i > t] \ , \quad \forall t \geq 0 \ ,$$

for each nonempty set $J \subset \{1,2,\dots,n\}$.

Details are given in Esary and Marshall (1974).

The characterization of the Marshall-Olkin multivariate distribution as a unique n-dimensional distribution satisfying the lack of memory property for $k = n$ with all $(n-1)$-dimensional marginals being Marshall-Olkin distributions, can be rephrased by saying that the lack of memory property holds for any k-dimensional marginal for $k = 1,2,\dots,n$. (The LMP for $k = 1$ yields the exponentionality of the univariate marginals.) The assumption of the lack of memory property for $k = 2,3,\dots,n$ implies that the minimum of two or more of the random variables $X_1, X_2, \dots, X_n$ with survival function $G(x_1,\dots,x_n)$ is exponential. Therefore for any $J \subset \{1,2,\dots,n\}$ containing more than one element we have a $\theta_j > 0$ such that $P[\min_{i\in J} X_i > t] = \exp[-\theta_J t]$ and the lack of memory property implies that

$$G(x_1,\dots,x_n) = \exp(-\theta_{12\dots n} x_{i_1}) G(x_1 - x_{i_1},\dots,0,\dots,x_n - x_{i_1})$$

for $0 \leq x_{i_1} \leq x_1 \leq \dots \leq x_n$ where the 0 appears in the i_1 position.

Successively applying this argument we obtain:

$$G(x_1,\dots,x_n) = \exp(-\theta_{12\dots n} x_{i_1}) \exp(\theta_{12\dots i_n}(x_{i_2} - x_{i_1}))\dots$$

$$\dots \exp(-\theta_{i_{n-1} i_n}(x_{i_{n-1}} - x_{i_n})) G_{i_n}(x_{i_n} - x_{i_{n-1}})$$

for $0 = x_{i_0} \leq x_{i_1} < \dots < x_{i_n}$, where G_{i_n} is the survival function of the i_n-th univariate marginal. (Note that so far we don't require the LMP for $k = 1$.) More compactly,

$$(14) \quad G(\underset{\sim}{x}) = \exp(-\sum_{k=1}^{n-1} \theta_{J_k}(x_{i_k} - x_{i_{k-1}})) G_{i_n}(x_{i_n} - x_{i_{n-1}})$$

for $0 = x_{i_0} \leq x_{i_1} \leq \dots \leq x_{i_n}$.

Moreover, since $G(\cdot)$ is a survival function, it can be easily verified, taking monotonicity into account that

$$\theta_{J_{n-1}} \leq \theta_{J_{n-2}} \leq \ldots \leq \theta_{J_1}$$

for all permutations $\{i_1, i_2, \ldots, i_n\}$ of $\{1,2,\ldots n\}$, where $J_1 = \{1,2,\ldots,n\}$ and $J_k = \{i_k, i_{k+1}, \ldots, i_n\}$ for $k = 2,\ldots,n$. It can also be shown, extending the corresponding argument presented by Marshall and Olkin (1967a) for the bivariate case (see, e.g. Block (1974) for a more detailed statement) that the requirement of absolute continuity of $G(\cdot)$ implies the equality

$$\theta_{i_{n-1} i_n} = \theta_{i_{n-k}} + \theta_{i_{n-k+1}} + \ldots + \theta_{i_n} \quad \text{for } k = 1,2,\ldots,n-1 .$$

Hence assuming the existence of exponential densities for the marginals (which is a consequence of the lack of memory property for $k = 1$) we arrive from (14) at the following expression for an absolute continuous $G(\cdot)$ satisfying the LMP for $k = 2,\ldots,n$ with exponential marginals:

$$G(\underset{\sim}{x}) = \exp(-\sum_{k=1}^{n} \theta_{J_k}(x_{i_k} - x_{i_{k-1}})) = \exp(-\sum_{k=0}^{n-1} (\theta_{J_k} - \theta_{J_{k+1}}) x_{i_k})$$

$$= \exp(-\sum_{k=1}^{n} \theta_{i_k} x_{i_k}) = \exp(-\sum_{k=1}^{n} \theta_k x_k) ,$$

i.e. $G(\cdot)$ is a product of its exponential marginals. In other words, we have proved

Theorem 5.3.4. The absolute continuity of the joint distribution coupled with the LMP for $k = 2,\ldots,n$ and the exponentionality of one-dimensional marginals lead to a multivariate exponential distribution with independent exponential components.

Therefore, a possible choice for an absolutely continuous MVE may be a distribution with one-dimensional marginals of the form of a mixture of exponential distributions. Such a distribution was constructed by Block and Basu (1974) in the bivariate case and by Block (1974) in the general multivariate case. There have also been a number of other ingenious attempts to modify and extend the multivariate Marshall-Olkin exponential distribution. Among these, the distributions proposed by Arnold (1974), (1975), Langberg, Proschan and Quinzi (1977), Esary and Marshall (1974), Marshall (1975) and Proschan and Sullo (1974) should be mentioned. Unfortunately, there are as yet no satisfactory characterization theorems for these distributions and this is one of the important and challenging problems in the theory and analysis of multivariate probabilistic models involving dependence among given marginals of a given type. Here we shall briefly describe a family of distributions suggested by Arnold (1974) in view of its close relation to a characterization of a univariate exponential distribution *not* included among the four equivalent characterizations discussed in Section 1.5.

Recall (Chapter 4) that Szántai (1971), Arnold (1973), Azlarov, Dzamirzaev and Sultanova (1972) have shown independently that if an integer valued random variable N has a geometric distribution while the variable $S_N = X_1 + X_2 + \ldots + X_N$,--where $\{X_i\}$ is a sequence of independent and identically distributed (i.i.d.) random variables with a common distribution function $F(\cdot)$ and N is independent of $\{X_j\}$-- is exponentially distributed, then each X_j is also exponentially distributed, $j = 1,2,\ldots$. Conversely, if S_N and X_j are exponential, then N is geometric. (In other words, geometric sums of exponential random variables are exponentially distributed.)

This result was extended by Arnold (1975) to the case of $\{X_1,X_2,\ldots\}$, a sequence of i.i.d. random variables and $\underset{\sim}{Y} = (Y_1,Y_2,\ldots,Y_n)$ having a multivariate geometric distribution with parameters $(k; p_1,\ldots,p_n)$ independent of $\{X_j\}_{j=1}^{\infty}$.

Theorem 5.3.5. Let $\underset{\sim}{S} = (S_1,\ldots,S_n)$ be an n dimensional random variable with components $S_i = \sum_{j=1}^{Y_i} X_j$. Then $S_1,S_2,\ldots,S_n$ are independent if and only if X_1 or $-X_1$ has a (univariate) exponential distribution.

Proof: We shall prove the *only if* part of the theorem. It is clear that without loss of generality we may take n = 2. Denote the characteristic function of X_1 by $\psi(t)$ and recall that the characteristic function of a univariate *geometric sum* (with parameter p) of i.i.d. variables with a common characteristic function $\psi(t)$ is

$$(15)\quad \frac{p\psi(t)}{1-(1-p)\psi(t)}$$

(cf. Section 4.4). Now from the structure of S, it follows that the characteristic function $\phi_{\underset{\sim}{S}}(\underset{\sim}{t})$ of $\underset{\sim}{S} = (S_1,S_2)$ can be expressed in terms of characteristic functions $\phi_{S_1}(t_1)$ and $\phi_{S_2}(t_2)$ as

$$(16)\quad \phi_{\underset{\sim}{S}}(\underset{\sim}{t}) = \frac{\psi(t_1+t_2)[p_1\phi_{S_2}(t_2) + p_2\phi_{S_1}(t_1)]}{1 - p_0\psi(t_1+t_2)},$$

where p_0,p_1,p_2 are the parameters of the multivariate geometrically distributed random variable $\underset{\sim}{Y} = (Y_1,Y_2)$. Define

$$(17)\quad \eta(t) = 1 - [\psi(t)]^{-1}$$

in a neighborhood of zero, R, say, in which $\psi(t)$ is bounded away from zero. Substituting (15) and (17) into (16) we easily obtain

$$(17a)\quad \phi_{\underset{\sim}{S}}(\underset{\sim}{t}) = \frac{p_1p_2(p_1+p_2 - \eta(t_1) - \eta(t_2))}{[p_1+p_2-\eta(t_1+t_2)][p_2-\eta(t_2)][p_1-\eta(t_1)]} .$$

In view of the independence of the components S_1 and S_2 we have, in particular,

$$(18)\quad \phi_{\underset{\sim}{S}}(\underset{\sim}{t}) = \phi_{\underset{\sim}{S}}(t_1,0)\phi_{\underset{\sim}{S}}(0,t_2)$$

or after substituting (17a) into the last equation we obtain that

$$\eta(t_1+t_2) = \eta(t_1) + \eta(t_2) \quad \text{for all } t_1 \text{ and } t_2$$

in the neighborhood R. The continuity of $\eta(t)$ in R yields $\eta(t) = at$, $\forall t \epsilon R$, where a is a complex number. In other words $\psi(t) = [1-at]^{-1}$ in the neighborhood R. Now since $\psi(t)$ is a characteristic function we deduce that $\overline{\psi(t)} = \psi(-t)$ which implies that $a = i\beta$ for some real number β. Therefore $\psi(t) = [1-i\beta t]^{-1}$ for all $t \epsilon R$ is recognized as a characteristic function of an exponential random variable with parameter $\lambda = |\beta|$ (for a positive β) or a "negative" exponential variable for a negative β. The direct part of this theorem, i.e. verification of the independence of S_i, $i = 1,\ldots,n$, under the exponential assumption involves some slightly tedious but straightforward calculations. □

This idea of successive multivariate geometric compounding has been exploited by Arnold (1975) to construct general classes of multivariate exponential distributions via the shock model extensions of the M-O multivariate exponential distributions. The hierarchal construction--at least in the bivariate case--leads to a recursive expression for the characteristic function (or a Laplace transform) for the k-th order bivariate vector $(X^{(k)},Y^{(k)})$ in terms of the corresponding function of the (k-1)-th order vector $(X^{(k-1)},Y^{(k-1)})$ as follows:

$$(19) \quad \phi_{X^{(k)},Y^{(k)}}(s,t) = p_{00}^{(k)}\phi_{X^{(k-1)},Y^{(k-1)}}(s,t)\phi_{X^{(k)},Y^{(k)}}(s,t) + p_{01}^{(k)}\phi_{X^{(k-1)},Y^{(k-1)}}(s,t)\phi_{X^{(k)}}(s) + p_{10}^{(k)}\phi_{X^{(k-1)},Y^{(k-1)}}(s,t)\phi_{Y^{(k)}}(t) + p_{11}^{(k)}\phi_{X^{(k-1)},Y^{(k-1)}}(s)$$

where it is assumed that the marginals of $(X^{(k)},Y^{(k)})$ are exponential with the means $\lambda\Pi_{i=1}^{k}(p_{10}^{(i)}+p_{11}^{(i)})^{-1}$, $\lambda\Pi_{i=1}^{k}(p_{01}^{(i)}+p_{11}^{(i)})$ respectively and $p_{00}^{(k)} + p_{01}^{(k)} + p_{10}^{(k)} + p_{11}^{(k)} = 1$ for each k and $(X^{(0)}, Y^{(0)}) = (X,X)$ where X is an exponential random variable with parameter λ.

This equation defines a bivariate exponential family which contains the Marshall-Olkin bivariate exponential as a special case with $p_{11}^{(2)} = 1$ and $k = 2$, and its multivariate analogues define the families of multivariate distributions which of course contain the Marshall-Olkin multivariate family and its various generalizations. An exponential choice of $\{X_1,X_2,\ldots\}$ and that of $(Y_1,\ldots,Y_n)$ obtained by minimization of a general multivariate geometric random variable leads to a Marshall-Olkin multivariate distribution for $(S_1,S_2,\ldots,S_n)$, where $S_i = \sum_{j=1}^{Y_i}X_j$, $i = 1,\ldots,n$, thus providing yet another characterization of this distribution.

A very similar device was used by Paulson (1975) to define a family of bivariate distributions by extending a characterizing equation of the univariate exponential distribution to the bivariate case.

Basically we are dealing with the characteristic functional equation of the form

(20) $\phi(t_1,t_2) = \psi(t_1,t_2)[p_{00} + p_{01}\,\phi(0,t_2) + p_{10}\phi(t_1,0) + p_{11}\phi(t_1,t_2)]$,

where ϕ and ψ are characteristic functions and the probabilities

$$p_{00} + p_{01} + p_{10} + p_{11} = 1 ,$$

with $p_{01} + p_{11} < 1$ and $p_{10} + p_{11} < 1$. As it was pointed out by Block (1977a), various choices for ϕ and ψ yield various models for bivariate distributions. For example, the choice

$$\psi(t_1,t_2) = [1 - i\theta(t_1+t_2)]^{-1}$$

and $p_{00} = 0$, $p_{01} = \theta/\theta_1$, $p_{10} = \theta/\theta_2$ and $p_{11} = 1-\theta/\theta_1-\theta/\theta_2$ results in the bivariate exponential with *independent marginals* given the characteristic function

$$\phi(t_1,t_2) = [(1-i\theta_1t_1)(1-i\theta_2t_2)]^{-1}$$

while the Marshall-Olkin bivariate exponential distribution corresponds to the same choice of $\psi(t_1,t_2) = [1-i\theta(t_1+t_2)]^{-1}$ but with $p_{01} = \theta\lambda_1$, $p_{10} = \theta\lambda_2$, $p_{00} = \theta\lambda_{12}$ and $p_{11} = 1-p_{00}-p_{01}-p_{10} = 1-\theta\lambda$, where $\lambda = \lambda_1+\lambda_2+\lambda_{12}$.

The Paulson bivariate distribution (1975) is obtained by choosing $\psi(t_1,t_2) = [(1-i\theta_1t_1)(1-i\theta_2t_2)]^{-1}$ (i.e. by taking $\psi(t_1,t_2)$ as a product of univariate exponential characteristic functions $\psi_1(t_1,0)$ and $\psi_2(0,t_2)$) and unrestricted probabilities p_{ij} $(i=0,1\ ,\ j=0,1)$ corresponding, for example, to probabilities taken by a 2×2 matrix valued random variable V with values in the

$$\left\{\begin{pmatrix} 0 & 0 \\ 0 & 0 \end{pmatrix}, \begin{pmatrix} 1 & 0 \\ 0 & 0 \end{pmatrix}, \begin{pmatrix} 0 & 0 \\ 0 & 1 \end{pmatrix}, \begin{pmatrix} 1 & 0 \\ 0 & 1 \end{pmatrix}\right\}$$

set. Using the notation $T = (t_1,t_2)$ for a real two-dimensional vector and $\psi(T) = E\{\exp(it_1U_1 + it_2U_2)\}$ for the characteristic function of the variable $\underset{\sim}{U} = (U_1,U_2)$ and $\phi(T) = E\{\exp(it_1X + it_2Y)\}$ for the characteristic function of (X,Y) equation (20) can be written as

$$\phi(T) = \psi(T)E[\phi(TV)] .$$

This is a formal bivariate analogue of the equation

$$\phi(t) = \psi(t)E[\phi(tV)]$$

which-- as it was shown by Paulson (1975)--- is the characteristic-functional equation satisfied under certain regularity conditions by the univariate exponential distribution and only by this distribution provided $\psi(t)$ is a characteristic function of an exponential distribution and $P[V=0] = 1-p$, $P[V=1] = p$; $0 < p < 1$.

In addition to the extensions of the multivariate M-O exponential distributions based on (multivariate) geometric compounding and on formal recursive equations for characteristic functions discussed above, a randomized extension of this distribution has been recently developed by Pickands III (1977). He gives several general characterization theorems for a class of multivariate exponential distributions.

Definition. The vector $(X_1,X_2,\ldots,X_n)$ or its distribution, is called exponential in the sense of Pickands if the survival function $G(x_1,x_2,\ldots,x_n)$ satisfies the relation

$$-t \log G(x_1/t, x_2/t,\ldots,x_n/t) = -\log G(x_1,x_2,\ldots,x_n)$$

for any vector $(x_1,x_2,\ldots,x_n)$ and any scalar $t, 0 < t < +\infty$.

Notice that the multivariate Marshall-Olkin distribution is an exponential distribution in the sense of Pickands, but there are several others as well. The following two theorems characterize Pickands' class of distributions.

Theorem 5.3.6. Let $\underset{\sim}{X}^{(1)}, \underset{\sim}{X}^{(2)},\ldots$ be independent, identically distributed m-dimensional random vectors. Then the common distribution of the $\underset{\sim}{X}^{(j)}$ is exponential in the sense of Pickands if, and only if, for every positive integer k, the vector

$$k \min(\underset{\sim}{X}^{(1)},\underset{\sim}{X}^{(2)},\ldots,\underset{\sim}{X}^{(k)})$$

has the same distribution as $\underset{\sim}{X}^{(1)}$. The minimum above is to be interpreted componentwise.

The reader will recognize the similarity of this characterization theorem to the univariate characterization of the exponential distribution presented in Section 1.5 as (P4*).

Theorem 5.3.7. Let each component X_j of the $\underset{\sim}{X} = (X_1,X_2,\ldots,X_n)$ be nonnegative. Then $\underset{\sim}{X}$ is exponentially distributed in the sense of Pickands if and only if, for every nonnegative vector $(z_1,z_2,\ldots,z_n)$ such that for at least one j, $z_j > 0$, the random variable

$$\min(X_1/z_1, X_2/z_2,\ldots,X_n/z_n)$$

has a univariate exponential distribution.

The class of Pickands' exponential distributions can be written in canonical form. Namely, the following result is valid:

Theorem 5.3.8. A survival function $G(x_1,x_2,\ldots,x_n)$ is exponential in the sense of Pickands if and only if there exists a finite measure $T(q_1,q_2,\ldots,q_n)$ on the Borel subsets of the unit simplex

$$S = \{(q_1,q_2,\ldots,q_n) \colon 0 \le q_j \le 1, \sum_{j=1}^{n} q_j = 1\}$$

such that

$$(21) \qquad -\log G(x_1,x_2,\ldots,x_n) = \int_S \max_{1\le j\le n} q_j x_j \, dT(q_1,q_2,\ldots,q_n) .$$

Notice that, for $n = 2$, if $T(q_1,q_2)$ is degenerate at a point, the above representation reduces to the bivariate Marshall-Olkin distribution. Hence the class of

Pickands' distributions can be viewed as a class of *randomized* Marshall-Olkin distributions. Such an extension is indeed very significant because Pickands' class and the representation (21) lead to a complete characterization of the multivariate extreme value distributions. See Galambos' (1978) book.

As J. Pickands III's (1977) results are as yet unpublished, we do not present further details here.

The above admittedly sketchy discussion is sufficient to show that we are now in the midst of burgeoning development in the area of multivariate exponential distributions and points up the existence of numerous as yet unexplored avenues in this direction.

Indeed, so far we have seen generalizations based mainly on the lack of memory property of the univariate exponential distribution and those derived from geometric compounding. It should be noted that equation (2a) and its multivariate extension are by no means the only possible meaningful form of the lack of memory property in several dimensions and even this natural approach has not been fully investigated as yet. In the next section we shall discuss some multivariate exponential distributions and their characterizations based on extending the concept of the hazard rate to the multivariate case.

5.4. Characterization of MVE distributions based on hazard rate properties.

As we have seen in (P1*), Section 1.5, the property of a constant hazard (or failure) rate is equivalent--among others--to the lack of memory property and completely characterizes the univariate exponential distribution.

In other words if the failure rate function

$$\frac{G(x+t)}{G(x)} \tag{22}$$

is constant in x--for each t-- where $G(x) > 0$ for all $x \geq 0$, then the distribution is exponential. Also if the distribution possesses density $f(\cdot)$ and if the hazard rate function

$$r(x) = \frac{f(x)}{G(x)} \tag{23}$$

is constant in x where $G(x) > 0$ for all $x \geq 0$ then the distribution is exponential.

Another way of stating this property is to say that in the univariate case only the exponential distributions are both increasing hazard rate (IHR) and decreasing hazard rate (DHR).

Turning to the multivariate case, we first observe that there are many possibilities of multivariate extensions and generalizations of the notions of failure and hazard rates. In fact, Marshall (1977) and Buchanan and Singpurwalla (1977) among others, list at least seven plausible definitions of multivariate distributions with monotone hazard rates and attempt to classify various multivariate exponential distributions based on extensions of the above stated property.

We shall briefly sketch Marshall's conditions and their consequences. As before $G(x_1,\ldots,x_n)$ denotes $P[X_1>x,\ldots,X_n>x]$ where $\underset{\sim}{X} = (X_1,\ldots,X_n)$ and $\underset{\sim}{x} = (x_1,\ldots,x_n)$.

Condition I. $\underset{\sim}{X}$ has an increasing hazard rate iff $G(\underset{\sim}{s}+\underset{\sim}{t})/G(\underset{\sim}{t})$ is decreasing (in the wide sense) in $\underset{\sim}{t} \geq 0$ for all $\underset{\sim}{s} \geq 0$ (and $G(\underset{\sim}{t}) > 0$ for all $\underset{\sim}{t} \geq \underset{\sim}{0}$). Similarly,

Condition I'. $\underset{\sim}{X}$ has a decreasing hazard rate if $G(\underset{\sim}{s}+\underset{\sim}{t})/G(\underset{\sim}{t})$ is increasing (in the wide sense) in $\underset{\sim}{t}$ for all $\underset{\sim}{s} \geq 0$ (and $G(\underset{\sim}{t}) > 0$ for all $\underset{\sim}{t} \geq \underset{\sim}{0}$).

Theorem 5.4.1. The only class of distributions for which both I and I' are valid (as we have essentially seen in Section 5.2 (cf. eq. (2) in that section)) is the class of multivariate exponential distributions with *independent* exponential marginals.

Denote $\underset{\sim}{e} = (1,\ldots,1)$. A somewhat less restrictive condition is

Condition II. $\underset{\sim}{X}$ has an increasing hazard rate if $G(\gamma\underset{\sim}{e}+\delta\underset{\sim}{e})/G(\delta\underset{\sim}{e})$ is decreasing in $\delta \geq 0$ such that $G(\delta\underset{\sim}{e}) > 0$ for all $\gamma \geq 0$ and this condition is also satisfied for all the marginals of all dimensions. (Here γ and δ are nonnegative scalars).

Analogously, to obtain Condition II' for DHR, replace increasing by decreasing and decreasing by increasing respectively in Condition II. (In Conditions II and II' we are dealing with the simultaneous survival of all the components to a common "horizon" time.) It is easy to verify that Condition II is equivalent to the condition that $\{\min_{i \in J} X_j\}$ is IHR [in the univariate sense based on equation (22)] for all nonempty subsets J of $(1,2,\ldots,n)$. Thus we have

Theorem 5.4.2. Conditions II and II' are both satisfied if and only if the multivariate $G(\cdot)$ has "exponential minima" in the sense defined in the previous section (i.e. the minimum of any nonempty subset of the corresponding random variables has an exponential distribution.)

Observe that the family of distributions with exponential minima is a rather wide family of distributions; it contains, for example, the Marshall-Olkin multivariate family.

A somewhat more stringent condition of IHR in the multivariate situation is given by

Condition III. $\underset{\sim}{X}$ has IHR if $G(\underset{\sim}{s}+\delta\underset{\sim}{e})/G(\delta\underset{\sim}{e})$ is decreasing in δ such that $G(\delta\underset{\sim}{e}) > 0$ for all vectors $\underset{\sim}{s} \geq 0$ and this condition is also satisfied for marginals of all dimensions.

Condition III' for DHR is obtained by interchanging "increasing" and "decreasing" in Condition III.

Using the arguments presented in the preceding section it is straightforward to show the validity of

Theorem 5.4.3. Conditions III and III' are both satisfied if and only if $G(\cdot)$ is the Marshall-Olkin multivariate exponential survival function.

As we shall see, it is also not difficult to verify the following

Theorem 5.4.4. Simultaneous fulfillment of conditions IV (and IV') of the form $G(\delta \underset{\sim}{e}+\underset{\sim}{t})/G(\underset{\sim}{t})$ decreasing (increasing) in $\underset{\sim}{t}$ together with these conditions on all the marginals for all $\delta \geq 0$ (and $G(\underset{\sim}{t}) > 0$) implies that $G(\cdot)$ is the Marshall-Olkin multivariate exponential survival function as well.

Remark: These conditions are termed multivariate increasing (decreasing) failure rates by Brindley and Thompson (1972).

Proof: If $\underset{\sim}{X}$ is a multivariate M-O distribution then

$$\frac{G(\underset{\sim}{t}-\delta\underset{\sim}{e})}{G(\underset{\sim}{t})} = \exp(-\delta(\sum_{1}^{n}\lambda_i + \sum_{i<j}\lambda_{ij}+\ldots+\lambda_{12\ldots n})$$

is constant in $\underset{\sim}{t}$ for each δ.

Conversely, if $X_1,\ldots,X_n$ satisfy both conditions IV and IV' and $G_k(\cdot)$ is the survival function of some set of them, then (in the obvious notation)

$$\frac{G_k(\underset{\sim}{t}_k+\delta\underset{\sim}{e}_k)}{G_k(\underset{\sim}{t}_k)} = c(\delta) \tag{24}$$

where the constant c depends on δ only. Now set $t_i = 0$, $i=1,\ldots,k$ and obtain from (24) that

$$c(\delta) = G_k(\delta\underset{\sim}{e}_k).$$

Whence

$$G_k(\underset{\sim}{t}_k+\delta\underset{\sim}{e}_k) = G(\underset{\sim}{t}_k)G_k(\delta\underset{\sim}{e}_k)$$

for any $\delta > 0$. We thus observe that the univariate marginals are exponential. The proof of the assertion is then completed using an induction argument of the type utilized above in Section 5.2. □

Actually the simultaneous fulfillment of the last conditions is equivalent to the lack of memory property. Other conditions studied by Marshall (1977) do not yield any new multivariate exponential distributions.

We have seen that multivariate distributions with independent exponential marginals and the M-O multivariate exponential distributions serve as the "boundary distributions" for certain definitions of multivariate IHR (DHR).

Extending the univariate notion of increasing (decreasing) hazard rate on the average, IHRA (DHRA)--see below--Buchanan and Singpurwalla (1975, 1977) introduced (among others) the concepts of multivariate increasing (decreasing) failure rate average--strong, MIFRA-S (MDFRA-S) and that of the multivariate increasing (decreasing) failure rate average--very weak conditional MIFRA-VWC (MDFRA-VWC) by requiring that in the case of MIFRA-S (MDFRA-S) we have

$$\left[\frac{G(\underset{\sim}{t}+\delta\underset{\sim}{e})}{G(\underset{\sim}{t})}\right]^{1/\delta} \downarrow (\uparrow)$$

in $\underset{\sim}{t}$ for all $\delta > 0$, all $\underset{\sim}{t} > 0$ and $G(\cdot) > 0$; and in the case of MIFRA-VWC (MDFRA-VWC):

$$\left[\frac{G(\gamma\underset{\sim}{e}+\delta\underset{\sim}{e})}{G(\gamma\underset{\sim}{e})}\right]^{1/\delta} \downarrow (\uparrow)$$

in γ for all γ and $\delta > 0$ and $G(\cdot) > 0$.

Utilizing the standard arguments involving induction presented in this and preceding sections and the univariate identities of the type $[\frac{G(t+\delta)}{G(t)}]^{1/\delta} \downarrow t$ implies that $1/\delta[\ell n\, G(t+\delta) - \ell n\, G(t)] \downarrow t$ implies that $\frac{\ell n\, G(t)}{t} \downarrow t$ (i.e.$-\ell n\, G(t)$ is a star-shaped function) we can directly verify the following theorem:

Theorem 5.4.5. The "boundary distributions" for the MIFRA-S and MDFRA-S classes of multivariate distributions defined on positive orthants as well as for MIFRA-VWC and MDFRA-VWC classes defined on the same region belong to the class of M-O multivariate exponential distributions.

Some related results were obtained by Esary and Marshall (1976) in an as yet unpublished report. Esary and Marshall utilize the univariate definition of IHRA distributions.

Definition. A random variable X with survival function $G(\cdot)$ is said to have an IHRA if $G(x) = 1$ for all $x < 0$ and if $[G(t)]^{1/t}$ is decreasing in $t > 0$. (If the hazard rate function $r(\cdot)$ as given by (23) exists, then this requirement is equivalent to the property that $t^{-1}\int_0^t r(x)dx$ is increasing in $t > 0$.)

A typical result obtained by Esary and Marshall is as follows:

Thoerem 5.4.6. Let $T_1, T_2, \ldots, T_n$ each be exponentially distributed. If there exist, for some k, independent IHRA random variables $X_1, \ldots, X_k$ such that for some non-empty subsets $J_i \subset \{1,2,\ldots,k\}$

$$T_i = \min_{j \in J_i} X_j \ , \quad i=1,\ldots,n \ ,$$

then $(T_1, \ldots, T_n)$ possess the multivariate M-O exponential distribution.

The Proof follows from the observation that under the assumption of the theorem, the "generating" random variables X_i, $i=1,\ldots,k$, are each exponentially distributed and from the standard argument used in Section 3 (compare with theorem 5.3.1). □

So far we have characterized multivariate distributions by stipulating simultaneously the multivariate IHR (or IFRA) and DHR (or DFRA) properties suitably defined, without defining the notion of multivariate hazard rate explicitly.

Another possibility is to require the constancy of an appropriately defined multivariate hazard rate. There are basically two definitions of this concept available in the literature. One is that of Basu (1971) which defines it as a scalar quantity

$$(25)\qquad r(\underset{\sim}{x}) = \frac{f(\underset{\sim}{x})}{G(\underset{\sim}{x})}$$

which he calls the *multivariate failure rate* where $f(\underset{\sim}{x})$ is the density of $G(\underset{\sim}{x})$. (If the density $f(\underset{\sim}{x})$ does not exist, we have various ways of extending the numerator in (22) to the multivariate case--this leads, as we have just seen, to various multivariate exponential distributions under the assumption of constancy of an extended ratio (22).)

Basu (1971) proved the following result:

Theorem 5.4.7. For bivariate distributions with a finite Laplace transform, if the function (25) is constant in x_1 and x_2 and possesses exponential marginals (i.e. the marginals have constant univariate hazard rates) then this bivariate distribution is the product of independent (and exponential) marginals.

Proof: We are to determine $f(x_1,x_2)$ such that

$$r(x_1,x_2) = \frac{f(x_1,x_2)}{G(x_1,x_2)} = \lambda$$

with

$$(26)\qquad G(x_1,0) = e^{-\lambda_1 x_1};\ G(0,x_2) = e^{-\lambda_2 x_2} \quad \text{and} \quad G(0,0) = 1\ .$$

This is equivalent to the second order partial differential equation

$$(27)\qquad \frac{\partial^2 G(x_1,x_2)}{\partial x_1\, \partial x_2} - \lambda G(x_1,x_2) = 0$$

with the boundary conditions (26).

Consider the Laplace transform corresponding to $r(x_1,x_2)$:

$$\phi(s,t) = \int_0^\infty \int_0^\infty e^{-(sx_1+tx_2)} r(x_1,x_2)dx_1 dx_2\ .$$

Integrating by parts (as it is often done in Chapters 1 and 2) with respect to x_1 and using $G(0,x_2) = e^{-\lambda_2 x_2}$, $x_2 \geq 0$, yields

$$\phi(s,t) = \frac{1}{s}\left\{\frac{1}{(t+\lambda_2)} + \iint e^{-(sx_1+tx_2)} \frac{\partial r(x_1,x_2)}{\partial x_1} dx_2 dx_1\right\}\ .$$

Next, integrating by parts with respect to x_2 and using the basic equation (27) and the boundary condition $G(x_1,0) = e^{-\lambda_1 x_1}$ we have

$$\phi(s,t) = \frac{1}{s(t+\lambda_2)} - \frac{\lambda_1}{st(s+\lambda_1)} + \frac{\lambda}{st}\phi(s,t)$$

which reduces to

$$(28)\qquad \phi(s,t) = \frac{st-\lambda_1\lambda_2}{(t+\lambda_2)(s+\lambda_1)(st-\lambda)}$$

and the Laplace transform of $f(x_1,x_2) = \lambda(G(x_1,x_2)$ denoted by $\psi(s,t)$ is therefore

$$(28a)\qquad \psi(s,t) = \frac{\lambda(st-\lambda_1\lambda_2)}{(t+\lambda_2)(s+\lambda_1)(st-\lambda)} \; .$$

However, it follows from the right hand side of (28) that in order that the Laplace transform $\phi(s,t)$ exist for all $s \geq 0$ and $t \geq 0$, the relation $\lambda = \lambda_1\lambda_2$ must be valid, and hence from (28a)

$$\psi(s,t) = \frac{\lambda_1}{(t+\lambda_2)} \times \frac{\lambda_2}{(s+\lambda_1)}$$

which is the Laplace transform of two independent exponential random variables. □

Remark: A similar proof can be obtained by observing that in the bivariate case $r(\underset{\sim}{x})$ as defined by (25) can be expressed under suitable assumptions on the survival function as:

$$r(x_1,x_2) = \frac{\partial}{\partial x_1} H(x_1,x_2) + \frac{\partial}{\partial x_2} H(x_1,x_2) + \frac{\partial^2}{\partial x_1 \partial x_2} H(x_1,x_2) \; ,$$

where $H(\cdot) = -\log G(\cdot)$ and solving the corresponding second order partial equation using for example Khinchine's method of generating functions prevalent in queueing theory problems.

This representation (not well known in the literature) can perhaps also be used to characterize classes of distributions with some other specified properties of $r(\underset{\sim}{x})$ (such as monotonicity) or by stipulating the explicit functional form of $r(\underset{\sim}{x})$ such as linearity.

As a Corollary we deduce that there is no absolutely continuous bivariate exponential distribution with constant hazard other than one with independent marginals. This is analogous to Block's result obtained in the preceding section in connection with the absolute continuous bivariate distribution satisfying the lack of memory property with exponential marginals.

It also follows from the proof above that a solution of equation (27) without the boundary conditions $G(x_1,0) = e^{-\lambda_1 x_1}$, $G(0,x_2) = e^{-\lambda_2 x_2}$ does not need to be independent exponential Indeed, the infinite mixture $G(x_1,x_2) = \sum_{i=1}^{\infty} e^{-\lambda_i x_1 - \mu_i x_2} p_i$, where $\lambda_i \mu_i = \lambda$, $i=1,\ldots,n$, and $\sum_1^{\infty} p_i = 1$ satisfies equation (27).

The last observation prompted Puri (1973) and Puri and Rubin (1975) to consider solutions of equation

$$\text{(25a)} \quad r(\underset{\sim}{x}) = \frac{f(\underset{\sim}{x})}{G(\underset{\sim}{x})} = \lambda$$

in a general finite multivariate case. Puri and Rubin (1975) obtain the following characterization of constancy of the scalar-valued multivariate failure rate:

Theorem 5.4.8. For a given $\lambda > 0$ the only absolutely continuous distributions satisfying (25a) are mixtures of exponential distributions given by

$$\text{(29)} \quad f(x_1,\ldots,x_n) = \lambda \int_0^\infty \ldots \int_0^\infty \exp[-\sum_1^n \lambda_j x_j] D(d\lambda_1,\ldots,d\lambda_n) \quad x_j \geq 0,\ j=1,\ldots,n ,$$

where the probability measure D is located on the set

$$A = [\ \prod_1^n \lambda_j = \lambda,\ \lambda_j > 0,\ j=1,\ldots,n]\ .$$

Remark: Note that the mixture $\lambda\sum_1^\infty p_i \exp[-\sum_{j=1}^n \lambda_{ij}x_j]$ with $\Pi_{j=1}^n \lambda_{ij}$ independent of i (=λ) and $\sum_1^\infty p_i = 1$ satisfies (25a).

A sketch of the proof. A rigorous proof of this theorem is somewhat technical, but the main ideas can be understood from the following sketch:

Equation (25a) is equivalent to

$$\lambda^{-1} f(x_1,\ldots,x_n) = \int_{x_1}^\infty \ldots \int_{x_n}^\infty f(y_1,\ldots,y_n)dy_1\ldots dy_n\ , \quad x_i > 0,\ i=1,\ldots,n\ .$$

Thus $f(\cdot)$ possesses derivatives of all orders with respect to $\underset{\sim}{x}$ and the derivatives vanish at infinity for any one of the arguments x_i, $i=1,\ldots,n$. Therefore

$$\text{(30)} \quad f(x_1,\ldots,x_n) = (-1)^n \lambda^{-1} \frac{\partial^n f(x_1,\ldots,x_n)}{\partial x_1,\ldots,\partial x_n}\ ; \quad x_i > 0,\ \forall i = 1,\ldots,n.$$

It also follows by an induction argument that the functions

$$\psi_{r_1,\ldots,r_n}(x_1,\ldots,x_n) = (-1)^{\sum_{i=1}^n r_i} \lambda^{-1} \frac{\partial^{\sum_{i=1}^n r_i} f(x_1,\ldots,x_n)}{\partial^{r_1} x_1 \ldots \partial^{r_n} x_n}$$

(i.e. partial derivatives of $f(\cdot)$ of various orders) which are nonnegative for all $x_i > 0$, $r = 0,1,2,\ldots$; $i=1,\ldots,n$ and also satisfy equation (30). The nonnegativity of $\psi_{r_1,\ldots,r_n}(\cdot)$ and the validity of equation (30) implies that $f(x_1,\ldots,x_n)$ is completely monotone and the representation theorem for completely monotone functions (see, e.g. Section 4.2 of Bochner (1960)) yields the existence of a probability measure D with respect to which the representation (29) is valid, while the constancy condition $r(\underset{\sim}{x}) = \lambda$ restricts the range of measures to the set $A = [\Pi_1^n \lambda_i = \lambda]$.

An alternative definition of a multivariate hazard function and that of a (*vector valued*) multivariate hazard rate was developed independently by Block (1973), Esary and Marshall (see Marshall (1975)) and Johnson and Kotz (1973a,b).

Definition. Let $\underset{\sim}{X} = (X_1,\ldots,X_n)$ be a random vector with probability distribution function $F_{\underset{\sim}{X}}(\underset{\sim}{x}) = P[\cap_1^n\{X_i \le x_i\}]$ and the survival function $G_{\underset{\sim}{X}}(\underset{\sim}{x}) = P[\cap_1^n\{X_i > x_i\}]$. The hazard function $H_{\underset{\sim}{X}}(\underset{\sim}{x}) \equiv -\log G_{\underset{\sim}{X}}(\underset{\sim}{x})$ and the vector-valued multivariate hazard rate (VMHR) is the vector

$$(31) \quad h_{\underset{\sim}{X}}(\underset{\sim}{x}) = \nabla H_{\underset{\sim}{X}}(\underset{\sim}{x}) = \left(\frac{\partial}{\partial x_1},\ldots,\frac{\partial}{\partial x_n}\right)\{-\log G_{\underset{\sim}{X}}(\underset{\sim}{x})\}$$

$$= \left(-\frac{\partial}{\partial x_1}\log G_{\underset{\sim}{X}}(\underset{\sim}{x}),\ldots,-\frac{\partial}{\partial x_n}\log G_{\underset{\sim}{X}}(\underset{\sim}{x})\right)$$

$$\equiv (h_{\underset{\sim}{X}}(\underset{\sim}{x})_1, h_{\underset{\sim}{X}}(\underset{\sim}{x})_2,\ldots,h_{\underset{\sim}{X}}(\underset{\sim}{x})_n) .$$

(Observe that $h_{\underset{\sim}{X}}(\underset{\sim}{x})_j = -\dfrac{\frac{\partial}{\partial x_j} G(\underset{\sim}{x})}{G(\underset{\sim}{x})}$ in the analogy of the univariate hazard rate

$h_X(x) = \dfrac{f_X(x)}{G_X(x)} = \dfrac{-\frac{d}{dx} G_X(x)}{G_X(x)}$ for a one-dimensional random variable X.)

Also note that $h_{\underset{\sim}{X}}(\underset{\sim}{x})_j$ which is the j-th component of the multivariate hazard rate corresponding to the multi-dimensional random variable $\underset{\sim}{X} = (X_1,\ldots,X_n)$ should be distinguished from the $h_{X_j}(x_j)$--the univariate hazard rate of the j-th component variable X_j.

The following result is immediate.

Theorem 5.4.9. If $X_1,\ldots,X_n$ are mutually independent then

$$G_{\underset{\sim}{X}}(\underset{\sim}{x}) = \prod_1^n G_{X_j}(x_j)$$

and

$$(32) \quad h_{\underset{\sim}{X}}(\underset{\sim}{x})_j = h_{X_j}(x_j) , \quad j = 1,\ldots,n .$$

Conversely, equality (32) implies independence of the component variables.

Now based on the definition (31), MIHR and MDHR distributions can be defined in various non-equivalent ways. (See Marshall (1977) and Buchanan and Singpurwalla (1977) quoted above and a recent report by Block and Savitz (1977) for an analysis of some of these definitions.) According to Johnson and Kotz (1975a) we have:

Definition. If for *all* values of $\underset{\sim}{x} = (x_1,\ldots,x_n)$ all the components of $h_{\underset{\sim}{X}}(\underset{\sim}{x})$ are increasing (decreasing) functions of the corresponding variable--i.e. $h_{\underset{\sim}{X}}(\underset{\sim}{x})_j$ is an increasing (decreasing) function of x_j for $j = 1,2,\ldots,n$--then the distribution is called (vector) multivariate IHR (DHR).

This definition allows us to distinguish between two types of constant hazard rates to be referred to below as a *strictly* constant hazard rate and a *locally* constant hazard rate respectively.

In the case of a strictly constant hazard rate we have

$$h_{\underset{\sim}{X}}(\underset{\sim}{x}) = \underset{\sim}{c} ,$$

with $\underset{\sim}{c} = (c_1, c_2, \ldots, c_n)$ an absolute constant with respect to *all* variables.

In the case of a locally constant rate each component of the multivariate hazard gradient is constant with respect to variation in the corresponding variable (i.e. $h_{\underset{\sim}{X}}(\underset{\sim}{x})_1$ does not depend on x_1, nor $h_{\underset{\sim}{X}}(\underset{\sim}{x})_2$ on x_2, etc.) but not with respect to the variation in the other variables.

Based on these definitions we have the following characterization theorem:

Theorem 5.4.10. The only multivariate distribution for which the multivariate hazard gradient is strictly constant is the exponential distribution with independent exponential marginals.

(Remark: This property can be shown to be equivalent to the *strong* lack of memory property as given by eq. (2) in Section 5.2 or to the simultaneous fulfillment of Marshall's (1977) conditions I and I' above.)

Proof: $h_{\underset{\sim}{X}}(\underset{\sim}{x}) = \underset{\sim}{c}$ implies that (whenever the hazard rate exists)

$$\frac{\partial \log G_{\underset{\sim}{X}}(\underset{\sim}{x})}{\partial x_j} = -c_j \qquad (j = 1,\ldots,n) .$$

Hence

$$G_{\underset{\sim}{X}}(\underset{\sim}{x}) = e^{-c_j x_j} g_j(x_1,\ldots,x_{j-1},x_{j+1},\ldots,x_n) \qquad (j = 1,\ldots,n) ,$$

whence $G_{\underset{\sim}{X}}(\underset{\sim}{x})$ is proportional to $\exp(-\sum_{j=1}^{n} c_j x_j)$. Substituting the boundary conditions satisfied by any multivariate survival function G, we deduce from the above relation that

$$G_{\underset{\sim}{X}}(\underset{\sim}{x}) = \exp(-\sum_{j=1}^{n} c_j x_j) . \qquad \square$$

The requirement of a locally constant multivariate hazard rate leads to a multivariate distribution which has not been encountered as yet in this chapter. To derive this result it is convenient to express the survival function in terms of the hazard rate using the fundamental equation due to Marshall (1975).

Recall that if $\phi(\underset{\sim}{x})$ is a function of n variables $x_1, x_2, \ldots, x_n$ with a continuous gradient $\nabla\phi(\underset{\sim}{x})$ on an open set $S \subset R^n$, then for any pair of points $\underset{\sim}{x}$, $\underset{\sim}{y} \in S$ and for every piecewise smooth curve Γ in S joining $\underset{\sim}{x}$ and $\underset{\sim}{y}$ we have the representation

$$\int_{\Gamma(\underset{\sim}{x},\underset{\sim}{y})} \nabla\phi(\underset{\sim}{t})d\underset{\sim}{t} = \phi(\underset{\sim}{y}) - \phi(\underset{\sim}{x}) ,$$

where the integral is interpreted as a line integral. Moreover, the value of the integral does not depend on the particular choice of the piecewise smooth curve Γ.

Suppose now that the support of the probability distribution of $\underset{\sim}{X} = (X_1, X_2, \ldots, X_n)$ is in the first quandrant of R^n, i.e. in $R_1^n = \{(x_1, x_2, \ldots, x_n), x_i \geq 0, i = 1, \ldots, n\}$ and let $h_{\underset{\sim}{X}}(\underset{\sim}{x}) = \nabla H_{\underset{\sim}{X}}(\underset{\sim}{x})$ exist and be continuous in an open set containing R_1^n. Choose an arbitrary path Γ which is orthogonal to the axis connecting point $\underset{\sim}{0} = (0, \ldots, 0)$ to a given point $\underset{\sim}{x} = (x_1, x_2, \ldots, x_n) \in R_1^n$. Since

$$\int_{\underset{\sim}{0}}^{\underset{\sim}{x}} h_{\underset{\sim}{X}}(\underset{\sim}{t})d\underset{\sim}{t} = \int_{\underset{\sim}{0}}^{\underset{\sim}{x}} \nabla H_{\underset{\sim}{X}}(\underset{\sim}{t})d\underset{\sim}{t} = H_{\underset{\sim}{X}}(\underset{\sim}{x}) - H_{\underset{\sim}{X}}(\underset{\sim}{0}) = H_{\underset{\sim}{X}}(\underset{\sim}{x}) = -\log G_{\underset{\sim}{X}}(\underset{\sim}{x})$$

and the line integral over Γ can be represented in the form

$$\int_{\underset{\sim}{0}}^{x} h_{X}(t)dt = \int_0^{x_1} h_{\underset{\sim}{X}}(t_1, 0, \ldots, 0)_1 dt_1 + \int_0^{x_2} h_{\underset{\sim}{X}}(x_1, t_2, 0, \ldots, 0)_2 dt_2 + \ldots + \int_0^{x_n} h_{\underset{\sim}{X}}(x_1, x_2, \ldots, x_{n-1}, t_n)_n dt_n$$

we have the representation

$$(33) \qquad G_{\underset{\sim}{X}}(\underset{\sim}{x}) = \exp - \left(\int_{\underset{\sim}{0}}^{\underset{\sim}{x}} h_{\underset{\sim}{X}}(\underset{\sim}{t})d\underset{\sim}{t} \right),$$

which is an extension of the well-known corresponding one-dimensional relationship, in which case the deduction is straightforward.

This representation allows us to prove the following theorem.

Theorem 5.4.11. The vector-valued multivariate hazard $h_{\underset{\sim}{X}}(\underset{\sim}{x})$ of a random vector $\underset{\sim}{X}$ is continuous and locally constant *if and only if* the joint distribution of $\underset{\sim}{X}$ is given by survival function

$$(34) \qquad G_{\underset{\sim}{X}}(\underset{\sim}{x}) = \exp - \left(\sum_{j=1}^{n} \theta_j x_j + \sum_{i<j} \theta_{ij} x_i x_j + \ldots + \theta_{1\ldots n} x_1 \cdots x_n \right)$$

with θ's ≥ 0 which is known as the Gumbel type I multivariate exponential distribution (see, e.g. Gumbel (1960) for the bivariate case).

Observe that hazard components of this distribution are *linear* in the variables and that the i-th component does depend on x_i, $i = 1, \ldots, n$.

For notational simplicity only, we shall present the proof for the case $n = 3$. The *if part* of the assertion is easily verified by direct differentiation.

Assume now that $h_{\underset{\sim}{X}}(x_1, x_2, x_3)$ is locally constant, i.e.

$$h_{\underset{\sim}{X}}(x_1, x_2, x_3) = (g(x_2, x_3), h(x_1, x_3), k(x_1, x_2)).$$

Integrating $h_{\underset{\sim}{X}}(\cdot)$ along the paths

$$(0,0,0) \to (x_1, 0, 0) \to (x_1, x_2, 0) \to (x_1, x_2, x_3)$$

and $(0,0,0) \to (0,x_2,0) \to (0,x_2,x_3) \to (x_1,x_2,x_3)$

we arrive at the equality

$$\int_0^{x_1} g(0,0)dt_1 + \int_0^{x_2} h(x_1,0)dt_2 + \int_0^{x_3} k(x_1,x_2)dt_3$$
$$= \int_0^{x_2} h(0,0)dt_2 + \int_0^{x_3} k(0,x_2)dt_3 + \int_0^{x_1} g(x_2,x_3)dt_1 ,$$

or, after integration we have:

$$(35) \quad (g(0,0) - g(x_2,x_3))x_1 + (h(x_1,0) - h(0,0))x_2 + (k(x_1,x_2) - k(0,x_2))x_3 = 0 .$$

The last equation shows that $k(\cdot,\cdot)$ is linear x_1, since the first summand on the left hand side is linear in x_1 and the second does not contain x_3. A similar argument yields that $g(\cdot,\cdot)$ is linear in x_3. We can thus set

$$(36) \quad g(x_2,x_3) = g_1(x_2) + g_2(x_2)x_3$$

and

$$(36a) \quad k(x_1,x_2) = k_1(x_2) + k_2(x_2)x_1 .$$

Substituting (36) and (36a) into (35) we obtain

$$(37) \quad (g_1(0) - g_1(x_2) - g_2(x_2)x_3)x_1 + (h(x_1,0) - h(0,0))x_2 + k_2(x_2)x_1x_3 = 0 .$$

Now observe from (35) that the functions $g(\cdot,\cdot)$ and $k(\cdot,\cdot)$ should be of the same degree in x_2, i.e. $g_i(\cdot)$ are of the same degree in x_2 as $k_i(\cdot)$ $(i = 1,2)$. On the other hand, (37) shows that $h(x_1,0)$ is linear in x_1. Hence setting $h(x_1,0) = h_2+h_{12}x_1$ and substituting into (37) we obtain

$$(g_1(0) - g_1(x_2) - g_2(x_2)x_3)x_1 + h_{12}x_1x_2 + k_2(x_2)x_1x_3 = 0$$

which implies that $g_1(\cdot)$ is linear x_2 and by the observation above so are $k_1(\cdot)$, $k_2(\cdot)$ and $g_2(\cdot)$. Hence $g(x_2,x_3)$ is a bilinear function of the form $g(x_2,x_3) = g_1 + g_{12}x_2 + (g_{21}+g_{22}x_2)x_3$ and similarly $k(x_1,x_2) = k_3 + k_{23}x_2 + (k_{13}+k_{123}x_2)x_1$, say. (Using some other path of integration (e.g. $(0,0,0) \to (0,0,x_3) \to (0,x_2,x_3) \to (x_1,x_2,x_3)$) we can easily show that $h(x_1,x_3)$ is of an analogous form. This, however, is not needed for our proof).

Substituting the obtained representations of $g(\cdot,\cdot)$, $k(\cdot,\cdot)$ and that of $h(x_1,0)$ into the fundamental representation (12) we have:

$$G_{\underset{\sim}{X}}(x_1,x_2,x_3) = \exp - \left(\int_0^{x_1} g(0,0)dt_1 + \int_0^{x_2} h(x_1,0)dt_2 + \int_0^{x_3} k(x_1,x_2)dt_3 \right)$$
$$= \exp - (g_1x_1 + h_2x_2 + k_3x_3 + h_{12}x_1x_2 + k_{23}x_2x_3$$
$$+ k_{13}x_1x_3 + k_{123}x_1x_2x_3)$$

as claimed. The conditions imposed on the marginals of $G_{\underset{\sim}{X}}(\cdot)$ of various dimensions imply the positivity of the constants g_i, h_i, h_{ij}, k_i, k_{ij}, and k_{123} appearing in the above equation.

Remarks. 1) The bivariate case corresponds to the original Gumbel's (1960) bivariate exponential distribution given in the standardized form by

(38) $G(x_1,x_2) = \exp(-x_1-x_2-\theta x_1 x_2)$; $\theta > 0,\ x_i \geq 0$

with the linear hazard rate components

$$h(x_1,x_2) = (1 + \theta x_2,\ 1 + \theta x_1) .$$

Conversely, any bivariate distribution with locally constant hazard rate

$$h(x_1,x_2) = (f_1(x_2),\ f_2(x_1))$$

yields distribution (38) or an appropriate linear transform of (X_1,X_2).

2) An alternative proof of this result can be found in Johnson and Kotz (1975a) (p. 498). (The proof presented in this chapter was suggested by C.S. Soong.)

Observe that the above arguments can be extended by imposing some other assumption on the behavior of hazard rate components. (However, the practicality of these types of characterization theorems depends upon the availability of efficient methods for estimating these components. See Ahmad and Lin (1977) in this connection.)

A result of this kind is implicit in Block (1974):

Theorem 5.4.12. If the hazard components $h_{\underset{\sim}{X}}(\underset{\sim}{x})_i$, $i = 1,\ldots,n$, are stationary in $x_1,\ldots,x_n$, and $G_{\underset{\sim}{X}}(\cdot)$ is absolutely continuous, then $G_{\underset{\sim}{X}}(\cdot)$ is LMP and conversely.

(Compare with Theorems 5.2.3 and 5.2.4 where, in addition, the exponentionality of the marginals is assumed. Note also that the hazard components for the multivariate Gumbel's exponential distribution (34) are not stationary.)

Proof: (W.A. Thompson, private communication).

Observe that for any $\delta > 0$,

$$\text{LMP} \iff \log G(x_1+\delta,\ldots,x_n+\delta) = \log G(x_1,\ldots,x_n) + \log G(\delta,\ldots,\delta) .$$

Therefore, directly by differentiation,

$$\text{LMP} \iff \text{grad} \log G(x_1+\delta,\ldots,x_n+\delta) = \text{grad} \log G(x_1,\ldots,x_n)$$

which shows that the hazard components are stationary.

On the other hand, if C is any smooth curve from 0 to $\underset{\sim}{x}$, then integration of these gradients along C yields LMP. □

Finally we note that in 1968 Bildikar and Patil introduced multivariate linear exponential-type distributions (obtained from Gumbel's (34) by setting

$\theta_{ij}, \theta_{ijk}$, etc. $\equiv 0$ and substituting in the expression for the density the multiplicative term by a function $h(x_1, x_2, \ldots, x_n)$ which does not involve θ_i $(i = 1, \ldots, n)$) and proved some characterization theorems based on the properties of pairwise and mutual independence. These results are discussed in detail in Johnson and Kotz (1972) and are therefore not repeated here. We note that these distributions seem to present a promising area for further investigations and generalizations and may provide some interesting characterization theorems related to dependence structures and hazard rates.

5.5. Summary.

In summary, it should be emphasized that the development of the theory of multivariate exponential distributions has proceeded in the last decade mainly along the lines of constructing multivariate models which are based on a suitably chosen specific physical model rather than on what would seem to be a more general approach --that of extending the characterizing properties of univariate exponential distributions. Characterization theorems--with occasional exceptions--are formulated and proved after a model has been properly described and analyzed and do not as yet serve, in general, as a stimulus and guiding light for new multivariate distributions. Moreover, again with few exceptions, there are no satisfactory characterization theorems for the several very ingenious multivariate distributions which were developed recently. This applies for example, to the generalization of the Freund-Weinman multivariate exponential distribution proposed by Block (1977) and that of Friday and Patil (1977). It is hoped that the unification and clarification of characterizations of univariate exponential distributions presented in the preceding chapters will aid researchers to extend these characterizations to the multivariate case resulting in sound models and variants of multivariate exponential distributions which are sufficiently wide and at the same time well-defined and the realm of their applicability can be conveniently checked.

To the best of the authors' knowledge, apart from a *rather well developed area of multivariate extreme-value distributions* (see, Galambos, 1978, Chapter 5 for a comprehensive and up-to-date treatment of this topic), there are as yet very few characterization theorems available for multivariate distributions formed by monotone transforms of multivariate exponential distributions except for direct exponential transforms on the marginals leading to multivariate Pareto distributions (Mardia, 1962) and power transforms leading to Weibull distributions mentioned above (Moeschberger, 1974)). We are aware of a number of as yet unpublished works investigating and characterizing multivariate Weibull distributions (with not necessarily Weibull marginals) based--for example--on the property of possessing Weibull minima after arbitrary scaling.

CHAPTER 6

MISCELLANEOUS RESULTS

6.1. Characterizations by statistical properties.

In previous chapters, we described a number of practical situations in which characterization theorems lead to a unique stochastic model for the respective "physical" model. In the present section, we present some characterization theorems which are based on statistical considerations. (The subject matter of this short chapter requires familiarity with some specialized concepts of mathematical statistics.)

The first group to be mentioned is that related to goodness of fit tests. We start with an example. Let $X_1, X_2, \ldots, X_n$ be independent observations on a random variable X with distribution function $F(x) = F(x,B)$, where B is an unknown parameter. We would like to test whether the distribution function $F(x,B)$ is exponential

(1) $$F(x,B) = 1 - \exp(-\frac{x}{B}) \ , \ x \geq 0 \ ,$$

where $B > 0$ is unspecified. In order to eliminate the parameter B, one can test the composite hypothesis

(2) H_0: $F(x,B)$ is given by (1)

by a simple one as follows. Define

(3) $$S_r = \sum_{i=1}^{r} X_i \ , \ 1 \leq r \leq n$$

and

(4) $$Y_{r:n-1} = \frac{S_r}{S_n} \ , \ 1 \leq r \leq n-1.$$

Now if H_0 is true then the joint distribution of $Y_{1:n-1}, Y_{2:n-1}, \ldots, Y_{n-1:n-1}$ is the same as that of the n-1 order statistics of n-1 independent and identically distributed random variables $Y_1, Y_2, \ldots, Y_{n-1}$, uniformly distributed on the unit interval (0,1). Therefore, H_0 of (1) can be replaced by the simple hypothesis

(5) H_0^*: $Y_{r:n-1}$, $1 \leq r \leq n-1$ of (4) are distributed as the order statistics of n-1 i.i.d. random variables uniformly distributed on the interval (0,1).

The hypothesis H_0^* can then be tested using the Kolmogorov-Smirnov test statistic

$$D_k^+ = \max_{1 \le r \le k} \left(\frac{r}{k} - Y_{r:k}\right) .$$

It is evident that the random variables $Y_{r:n-1}$ of (4) are not affected by a scale parameter B. We can ask from a purely mathematical point of view whether the hypotheses H_0 and H_0^* are equivalent? In other words, we are interested in the question whether H_0 is true if, and only if, H_0^* is true. When the answer is affirmative, we evidently obtain a characterization theorem for the exponential distribution (1). The Canadian group of mathematicians, M. Csörgö, V. Seshadri, M.A. Stephens and M. Yalovsky, has shown that if $n \ge 3$, $P(X \ge 0) = 1$ and $F(x,B)$ admits density (with respect to Lebesgue measure), then H_0 and H_0^* are indeed equivalent (see Csörgö, Seshadri and Yalovsky (1974 and 1975), Csörgö and Seshadri (1970) and (1971) and Seshadri, Csörgö and Stephens (1969)). In these same papers, the authors recommend a new set of random variables, based on the original observations, in terms of which a composite hypothesis can again be replaced by a simple one, when an unknown translation parameter is present. More precisely, if the observations $X_1, X_2, \ldots, X_n$ are such that $X_i > A$, where A is a real number and if one would like to test H_0 for the sequence $X_j - a$, where now both $a \le A$ and $B > 0$ are unknown, the following transformations are recommended. Let $X_{r:n}$ be the r-th order statistic of the X_j. Set

$$\Delta_{i:n} = (n+1-i)(X_{i:n} - X_{i-1:n}) ,$$

where $1 \le i \le n$ and $X_{0:n} = a$. Finally, define

$$S_r = \sum_{i=2}^{r} \Delta_{i:n} \text{ and } Y_{r:n-2} = \frac{S_r}{S_n} ,$$

$r = 2,3,\ldots,n-1$ $(n \ge 4)$. In view of Theorem 3.1.1, if the common distribution of $X_j - a$, $j \ge 1$, is $F(x,B)$ given by (1), then the differences $\Delta_{i:n}$ are also exponential and their distribution does not depend on a. Consequently, the distribution of $Y_{r:n-2}$, $2 \le r \le n-1$, is the same as that of the order statistics of n-2 independent random variables uniformly distributed over the interval (0,1), whenever H_0 is true for the transformed variables X_j-a. The advantage of transforming X_j to the sequence $Y_{r:n-2}$, $r \ge 2$, is that their distribution is free of both a and B. (The original recommendation given in the above mentioned papers is different from the one described above in that these authors define S_r as the sum of $\Delta_{i:n}$, where i runs from one to r. However, $\Delta_{1:n}$ requires the knowledge of a, which makes the whole transformation unnecessary).

The question again arises whether the joint distribution of the $Y_{r:n-2}$ characterizes the common distribution of the variables $(X_j-a)/B$, where a and B are unspecified real numbers? The answer to this question is still an open problem, if we discount trivialities. The emphasis here is to use only the $Y_{r:n-2}$ in a characterization theorem, i.e., not to make separate assumptions on the sequence $\Delta_{i:n}$. Namely, characterizations based on the $\Delta_{i:n}$ have been explored in Chapter 3. It would indeed be very instructive and useful to settle this question. Our conjecture is that the answer is negative to this problem.

In connection with applications of characterization theorems in tests for exponentiality, see also the paper by Y.H. Wang and S.A. Chang (1977).

We also mention, briefly, the following important area of current research.

In recent years a number of characterizations of distributions by optimality properties of certain statistics within a given family have received special attention.

One of the main results in this area is the discovery by Bondesson (1973) that the arithmetic mean $\overline{X}$ is a Uniformly Minimum Variance (UMV) estimator of the location in the location-scale type family $F(\frac{x-a}{B})$ if and only if the underlying population is gamma. Another result along these lines is due to Klebanov and Melamed (1976) and deals with Bayesian estimators.

Although there are no immediate practical applications of these characterization theorems they seem to be important for the following reasons:

1) They have led to discoveries of new distributions or to new cases for which the considered properties are valid (an example is the von Mises distribution which shares some of these properties with the normal distribution).
2) Characterizations of distributions through optimality properties of some *invariant* estimators has led to a better understanding of the relation between estimation characterizations and characterizations through independence of certain statistics and thus to a better insight in the relation between different properties for particular distributions.
3) The fact that $\overline{X}$ is a UMV estimator (or an MSE estimator) of the "population mean" for a location parameter family in the case of a normal distribution *only* shows that this distribution is - in a certain sense - the most random one. This property is closely connected with the maximum entropy and minimum Fisher information characterizations of the normal distribution and renders the central limit theorem to become "highly" plausible.

For the exponential distribution a characterization along these lines is given by the following result of Cifarelli and Regazzini (1976): Within the scale family $F(x/B)$ possessing densities and satisfying certain rather cumbersome "regularity conditions," the estimator $\overline{X}$ is unbiased and efficient (i.e. results in an equality in the Cramèr-Rao inequality) for B if, and only if, the density

$f(x;B)$ is of the form:

$$f(x;B) = \begin{cases} 0 \,, & x < 0 \\ \frac{1}{B} e^{-x/B} \,, & x \geq 0 \,. \end{cases}$$

This characterization is closely related to the characterization via maximum likelihood estimators due to Teicher (1961) discussed in detail in Kagan, Linnik and Rao's (1975) book.

The relation between independence and sufficiency is a topic of significant importance in statistical methodology; a short note by D. Basu (in 1958) stimulated much of the research in this area. On the other hand, the independence between the sample mean $\overline{X}$ and sample variance S is the cornerstone characterization of the normal distribution as it was discussed in the Introduction.

In 1970 Kelker and Mattes showed that within location and scale parameter families, $(\overline{X},S)$ is a *sufficient* statistic if, and only if, the family is normal. (As it was pointed out by Bondesson (1977) it is necessary to assume that the distribution F is continuous for their proof.)

In 1977 Bondesson showed - interalia - that the property that $(\overline{X},X_{1:n})$ (where $X_{1:n}$ is the first order statistic) is a sufficient statistic within a location-scale continuous family implies that the underlying population is an exponential distribution (additional requirements that F possess a finite mean and that the sample size $n \geq 6$ are required for Bondesson's proof). The converse of this result is obtained in the 1954 classical paper by Epstein and Sobel.

Gokhale's (1975) result which is implicit in Reza (1961) is as follows:

Among all continuous distributions with given mean and $(0,\infty)$ as support, the exponential distributions $F(x;B) = 1-\exp(-x/B)$, $0 < B < \infty$, $0 < x < +\infty$ maximize the entropy $-\int f(x)[\ell n\, f(x)]dx = -E[\ell n\, f(X)]$, where $f(x) = F'(x)$.

Proof: The assumptions lead directly to an equation of the form

$$-(1 + \ell nf) + \mu + \lambda x = 0$$

for some μ and λ. Thus ℓnf is linear in $\underline{x}$ which yields that $f(\cdot)$ is indeed an exponential density.

6.2. Records

Let $X_1,X_2,\ldots$ be independent observations on a random variable X whose distribution function F(x) is continuous. When a new "high value" occurs in the sequence X_j, we call it a record. More mathematically, X_1 is a record. Also, X_j is a record if $X_t < X_j$ for all $t < j$. Let $R_1 < R_2 < \ldots$ be the successive records in the *infinite* sequence $X_1,X_2,\ldots$. Evidently, $R_j = X_{m(j)}$ for some random $m(j) \geq j$

($m(1) = 1$). It is also evident that the distribution of $m(j)$ does not depend on $F(x)$, since monotonic transformations preserve inequalities of the form $X_t < X_j$, on which the definition of records is based. However, the distribution, and even the dependence structure, of the sequence R_j, $j \geq 1$, is dependent on the parent distribution $F(x)$. In this section, we shall quote two results related to distributional properties of the sequence R_j, which characterize the exponential distribution as $F(x)$. For details on the general theory of records and for further references, see Chapter 6 in the book by Galambos (1978).

It can be seen by an easy calculation that, if $F(x) = 1 - \exp(-bx)$, $x \geq 0$, then the differences $M_j = R_j - R_{j-1}$, $j \geq 1$, ($R_0 = 0$), are independent random variables whose common distribution function is $F(x)$ itself. In her basic paper, M.N. Tata (1969) proved a converse to this statement. We formulate this as

Theorem 6.2.1. Let X_n, $n \geq 1$, be independent random variables with common absolutely continuous distribution function $F(x)$. Then $F(x)$ is exponential if, and only if, $R_1 = X_1$ and $R_2 - R_1$ are independent.

Remark. The conclusion remains to hold if we relax the assumption of absolute continuity to continuity. We kept Tata's original formulation here because her proof shows, without substantial calculations, that the assumption of the theorem is in fact reduced to the validity of the lack of memory property.

Proof: The joint density $g(x,y)$ of R_1 and R_2 can be computed by observing that ($R_1 = x$, $R_2 = y$) is the union of events of the form

$$\{X_1 = x,\ X_j < x,\ 2 \leq j \leq m-1,\ X_m = y\}\ ,\quad m = 2, \dots ,$$

where $x < y$. Hence

$$g(x,y) = \sum_{m=2}^{+\infty} f(x) F^{m-2}(x) f(y)\ ,$$

where $x < y$ and $f(x) = F'(x)$. We thus have

$$g(x,y) = \frac{f(x)}{1-F(x)} f(y)\ ,\ x < y,$$

and $g(x,y) = 0$ for $x \geq y$. This yields that the density of $(R_1, R_2 - R_1)$ equals

$$(6) \quad h(x,z) = \frac{f(x) f(x+z)}{1-F(x)}\ ,\quad -\infty < x < +\infty\ ,\ z \geq 0\ .$$

Now, if the distribution $F(x)$ is exponential, then (6) reduces to a product of two functions, where one is a function of x and the other one of z only.

Conversely, if R_1 and $R_2 - R_1$ are independent then

$$\frac{f(x)F(x+z)}{1-F(x)} = f(x)s(z) ,$$

where $s(z)$ is the density of R_2-R_1. Simplifying by cancelling $f(x)$ and integrating with respect to z, we obtain

(7) $$\frac{F(u+x)-F(x)}{1-F(x)} = S(u) .$$

Since the functions above are continuous, letting $x \to 0$ yields that

$$S(u) = \frac{F(u)-F(0)}{1-F(0)} = 1 - \frac{1-F(u)}{1-F(0)} .$$

Hence, (7) reduces to

(8) $$G(u+x)G(0) = G(u)G(x) ,$$

where $G(u) = 1 - F(u)$. This is the lack of memory equation for $G(z)/G(0)$ (since $F(x)$ is absolutely continuous and thus continuous, $G(0) = 0$ can not be valid in (8)). Hence, $G(z)/G(0)$ is exponential, as claimed. □

Analogously to the case of order statistics (see Chapter 3), Theorem 6.2.1 has a variant where the independence of R_1 and R_2-R_1 is replaced by a constant regression of R_1 on R_2-R_1. An even more general result is valid, due to H.N. Nagaraja (1977). With previous notations, we have

Theorem 6.2.2. Let $h(x)$ be a strictly increasing function on an interval $[a,b]$ such that $F(a) = 0$ and $F(b) = 1$ and

$$\int_a^b h(x)dF(x)$$

is finite. Let $K(y) = E[h(R_2)|R_1 = y]$. Assume that $K(y)$ is a nondecreasing function for almost all y (with respect to F). Then $K(y)$ uniquely determines the distribution function $F(x)$.

The actual proof is simple and it can be proved by the same straightforward method that we utilized in Section 2.3.

Nagaraja points out by an example that the theorem fails to hold without the monotonicity assumption on $K(y)$.

6.3. Further results in terms of order statistics.

We have mentioned in Section 3.3 that the following general problem was settled by Ferguson (1967) and Kemperman (1971).

Let X_1 and X_2 be independent random variables with distribution functions

$F_1(x)$ and $F_2(x)$, respectively. Let $W = \min(X_1,X_2)$ and $R = |X_1-X_2| = \max(X_1,X_2) - W$. What can be said about F_1 and F_2 if W and R are independent?

Ferguson (1967) completely described the discrete pairs (F_1,F_2) which are solutions to the above problem. He also gave a general class of continuous pairs (F_1,F_2), which show that some nonexponential distributions are also included in the set of possible solutions.

For easier reference, let S be the set of pairs (F_1,F_2) of distribution functions for which W and R are independent. Kemperman (1971) established the following important and elegant results.

Theorem 6.3.1. The set S is closed under convergence in distribution.

Theorem 6.3.2. All continuous distributions (F_1,F_2) belonging to S are limits of four basic discrete classes of S.

Theorem 6.3.3. There are only three types of continuous pairs in the set S.

The four discrete classes referred to in Theorem 6.3.2 are those found by Ferguson. These classes may be called geometric type since their supports are the nonnegative integers and

$$P(X_i = k) = \begin{cases} c_{1,i}p^k & \text{if } k \in A, \\ d_{1,i}s^k & \text{otherwise} , \end{cases}$$

where A is a well defined subset of the nonnegative integers. Similarly, the new distributions obtained in Theorems 6.3.2 and 6.3.3 as limits of the above geometric types may be called exponential types because their functional forms involve solely exponential functions.

The additional possible members of S are two degenerate classes.

The proof of these results by Kemperman is quite technical and yet very neat. Of course, it cannot be represented without reproducing the whole paper which evidently is not our aim.

Kemperman (1971) announces a number of other results which warn that it is dangerous to try to imitate the exponential distribution by the (discrete) geometric distribution. Although the relation between these two distributions is indeed very strong, most results established for the exponential distribution in Chapter 3 in terms of order statistics fail to be valid for the geometric distribution when the sample size exceeds two.

On the other hand, the method of proof developed by Kemperman (1971) indicates that the role of discrete distributions in general is very basic in that discrete distributions can be used to find all continuous solutions of a particular problem. Hence, the theory of order statistics of discrete distributions deserves a more significant amount of attention and development than it has received in the current literature. In particular, it would be interesting and challenging to extend several results obtained by Galambos (1975b).

We take this opportunity to draw attention to another aspect of the relationship between exponential and geometric distributions: the so-called "clocking property," which is discussed in the recent note by Hawkins and Kotz (1976). (In this note some unsolved problems of practical interest concerning the relationship between these two distributions are posed).

We have emphasized in all previous chapters that characterization theorems are important and powerful tools for building stochastic models. This fact is not always well understood. In several instances, scientists start with models which seem to be very general. However, if one looks at the underlying assumptions it may turn out that only a few, or often a single, model can satisfy those assumptions. Several queueing models fall into this category. Another case in point occurs in the theory of competing risks. H.A. David (1970) discusses the consequences of some assumptions on the ratios of the hazard rates of some hypothetical random life lengths which would apply if the other risks were not present. It turns out that the underlying distributions are strongly related to the so-called extreme value distributions and, in particular, the assumptions of the discussed model are closely linked to a characterization theorem by Sethuraman (1965) (a version of which is formulated as Theorem 3.2.1 in Chapter 3).

We conclude this section by posing a question of theoretical interest. In connection with the central limit theorem, J. Sethuraman (1977) proved that if $X_1,X_2,\ldots,X_n$ are independent random variables with common distribution function $F(x)$, then the limit, (whenever it exists)

$$\lim_{n=+\infty} \frac{1}{n} \log P(X_1+X_2+\ldots+X_n \geq na) = d_F(a)$$

uniquely determines F. That is, a large deviation probability has a unique order of magnitude for a given parent distribution. The question is whether such a characteristic property could be found within the theory of weak convergence of extremes? More specifically, if $F(x)$ is such that there exist constants a_n and $b_n > 0$ for which

$$(9) \quad P(Z_n < a_n + b_n x) \to \exp(-e^{-x}) ,$$

where

$$Z_n = \max(X_1,X_2,\ldots,X_n) ,$$

then does the magnitude of

$$\log P(Z_n \geq a_n y)$$

characterize F(x)? For necessary and sufficient conditions concerning the validity of (9), see the first four sections of Chapter 2 in the book by Galambos (1978).

6.4. Some additional characterizations.

We first describe a result related to our basic properties in Chapter 1 and then we formulate yet another problem which would be interesting to settle.

Shantaram and Harkness (1972) discuss the class of distribution functions F of nonnegative random variables (i.e., $F(x) = 0$ for $x < 0$) defined by the integral-functional equation of the form

$$(10) \quad F(x) = b^{-1} \int_0^{tx} [1-F(u)]du \ , \quad x > 0 \ ,$$

with $t \geq 1$.

Equation (10) arises as the limit of a sequence $\{F_n(x)\}$ of absolutely continuous distribution functions defined recursively

$$(11) \quad F_n(x) = \mu_{n-1} \int_0^{x} [1-F_{n-1}(y)]dy \ , \quad x > 0 \ ,$$

with

$$F_1(x) = \mu \int_0^{x} [1-F_0(y)]dy \ , \quad x > 0$$

where F_0 is the (initial) distribution function of a nonnegative random variable X, $\mu = E(X)$ and $\mu_{n-1} = \int_0^{\infty}[1-F_{n-1}(y)]dy$ are the means of the successively generated random variables. The reader will definitely note the close resemblance between (10) and property (P2) discussed in Chapter 1 (see in particular Table 1 on p. 17). (Note also the proofs of Theorem 1.3.1 in Chapter 1 which are iterative in their nature.)

From these observations it is clear that the unique solution of (10) for $t = 1$ is the exponential distribution $F(x) = 1 - e^{-x/B}$ for $x > 0$.

For $t > 1$ there exist infinitely many distribution functions satisfying equation (10). In particular, the distribution function F(x) of the product $Z = UV$, where U and V are independent with densities f(u) and g(v) given by

$$f(u) = Be^{-1/2}\exp(-ue^{1/2}/B) \ ,$$

an exponential density with parameter $Be^{-1/2}$, and

$$g(v) = v^{-1}(2\pi \log t)^{1/2}\exp[-(\log v)^2/2 \log t] \ ,$$

where $v > 0$, is a solution of (10).

Moreover, it is easily seen that, along with $g(v)$, any density of the form

$$[1 + \alpha \sin(2\pi \log v)]g(v) , \quad |\alpha| < 1 ,$$

can serve as the density of V in $Z = UV$ and the distribution of Z will continue to satisfy (10). This leads to the following stability problem.

It would be interesting, both from a theoretical and a practical point of view, to investigate the stability of the characterizing property (P2) of the exponential distribution in terms of equation (10) (see also Section 2.2). Namely, by setting $t = 1+\varepsilon$ in (10), what is the optimal bound in terms of ε on the maximal deviation between the corresponding exponential distribution and the "worst" representation in the class of solutions of equation (10)?

In concluding we wish to draw attention to the paper by G.P. Patil and V. Seshadri (1964), which has had a significant influence on the early development of the theory of characterizations. This influence was especially evident at the Calgary Conference on Characterizations of Distributions already mentioned in the previous chapters. (Their paper was one of the most frequently referred to in the *Proceedings* of that conference.) A newer result, somewhat related to the general approach of Patil and Seshadri, is obtained in the article by S. Talwalker (1977), whose work is also strongly related to the subject matter treated in Chapter 2.

Finally, we would like to mention the paper by H. Pisarewska (1975) who characterizes the exponential distribution by a general type of independence criterion. The original assumptions are quite involved but it turns out that they are quite easily reduced to the constancy of the hazard rates (cf. property (P1) in Table 1 on page 17) of the variables under consideration. This work was motivated by the system of equations (9.5.7) on p. 345 of Fisz's book (1963) which deals with the characterization of the normal distribution based on independence of the mean and the standard deviation (cf. Section 1.1). The difference between closely related characterizing equations which lead to an exponential distribution on one hand and to the normal distribution on the other is clearly exemplified in this paper.

ADDENDUM

After the manuscript has been submitted for publication, we received the December 1977 issue of the *Journal of Applied Probability* which contains the paper by Berk, R.H. (1977), Characterizations via conditional distributions (Vol. 14, 806-816). The author of this paper extends both the original result of Patil and Seshadri (1964) and its order statistics variant by Galambos (1975a) in the same manner as the lack of memory property is weakened in Section 2.1 (p. 23).

BIBLIOGRAPHY

Aczel, J. (1966) *Lectures on Functional Equations and their Applications*, Academic Press, New York.

Aczel, J. (1975) General solution of a functional equation connected with a characterization of statistical distributions, *Statist. Distrib. in Scientific Work*, Vol. 3, 45-55 (G.P. Patil, S. Kotz, J. Ord, eds.), Dordrecht-Boston; D. Reidel.

Ahmad, I.A. and Lin, P.E. (1977) Non-parametric estimation of a vector-valued bivariate rate, *Ann. Math. Statist.*, 5, 1027-1038.

Ahsanullah, M. (1972) Characterization of exponential distribution (Abstract). *Bull. Inst. Math. Statist.* 1, 174.

Ahsanullah, M. (1975) A characterization of the exponential distribution, *Statist. Distrib. in Scientific Work*, Vol. 3, 131-135 (G.P. Patil, S. Kotz, J. Ord, eds.), Dordrecht-Boston; D. Reidel.

Ahsanullah, M. (1976) On a characterization of the exponential distribution by order statistics, *J. Appl. Prob.*, 13, 818-822.

Ahsanullah, M. (1977a) A characteristic property of the exponential distribution, *Ann. Statist.*, 5, 580-582.

Ahsanullah, M. (1977b) A characterization of the exponential distribution by spacings. To appear.

Ahsanullah, M. and Rahman, M. (1972) A characterization of the exponential distribution, *J. Appl. Prob.*, 9, 457-461.

Ali, M.M. (1976) An alternative proof of order statistics moment problem, *Can. J. of Statist.*, 4, No. 1, 151-153.

Arnold, B.C. (1967) A note on multivariate distributions with specified marginals, *J. Amer. Stat. Assoc.*, 62, 1460-1461.

Arnold, B.C. (1971) *Two characterizations of the Exponential Distribution using Order Statistics*, unpublished manuscript. Iowa State University, Ames, Iowa.

Arnold, B.C. (1973) Some characterizations of the exponential distribution by geometric compounding, *SIAM J. Appl. Math.*, 24, 242-244.

Arnold, B.C. (1975) A characterization of the exponential distribution by multivariate geometric compounding, *Sankhyā, Ser. A*, 37, 164-173.

Arnold, B.C. and Isaacson, D. (1976) On solutions to $\min(X,Y) \overset{d}{=} aX$ and $\min(X,Y) \overset{d}{=} aX \overset{d}{=} bY$. *Z. Wahrschein. verw. Gebiete,* 35, 115-119.

Arnold, B.C. and Meeden, G. (1975) Characterizations of distributions by sets of moments of order statistics, *Ann. Statist.*, 3, 754-758.

Arnold, B.C. and Meeden, G. (1976) Another characterization of the uniform distribution, *Austr. J. of Statist.*, 18, 173-175.

Azlarov, T.A., Dzamirzaev, A.A. and Sultanova, M.M. (1972) Characterization properties of the exponential distribution and their stability, *Sluchain. Proc. i Statist. Vyvody,* 2, Tashkent, Fan, 10-19 (in Russian).

Azlarov, T.A. (1972) Stability of characterization properties of the exponential distribution, *Litovs. Mat. Sbornik,* 12, No. 2, 5-9 (in Russian).

Basu, D. (1958) On statistics independent of sufficient statistics, *Sankhyā,* 20, 223-226.

Basu, A.P. (1965) On characterizing the exponential distribution by order statistics, *Ann. Inst. Statist. Math.*, 17, 93-96.

Basu, A.P. (1971) Bivariate failure rate, *J. Amer. Statist. Assoc.*, 66, 103-104.

Basu, A.P. and Block, H.W. (1975) On characterizing univariate and multivariate ex ponential distributions with applications, *Statist. Distrib. in Scientific Work*, Vol. 3, 399-491 (G.P. Patil, S. Kotz, J. Ord, eds.) Dordrecht-Boston; D. Reidel.

Beg, M.I. and Kirmani, S.N.U.A. (1974) On a characterization of exponential and related distributions, *Austr. J. Statist.*, 16, 163-166; correction 18, No. 1, p. 85.

Bernstein, S.N. (1941) Sur une propriété characteristique de la loi de Gauss, *Trans. Leningrad Polytechn. Inst.* 3, 21-22.

Bildikar, S. and Patil, G.P. (1968) Multivariate exponential type distributions, *Ann. Math. Statist.*, 39, 1316-1326.

Block, H.W. (1973) *Monotone hazard and failure rates for absolutely continuous multivariate distributions*, Dept. of Math. Research Report #73-20, University of Pittsburgh, Pittsburgh, Pa.

Block, H.W. (1974) *Constant multivariate hazard rate and continuous multivariate extensions*, Dept. of Math. Research Report #74-12, University of Pittsburgh, Pittsburgh, Pa.

Block, H.W. (1975) Continuous multivariate exponential extensions, *Reliability and Fault Tree Analysis*, 285-306, (R. Barlow, J. Fussell and N. Singpurwalla, eds.) SIAM, Philadelphia.

Block, H.W. (1977) A characterization of a bivariate exponential distribution, *Ann. Statist.*, 5, 808-812.

Block, H.W. (1977a) A family of bivariate life distributions, *Theory and Applications of Reliability*, Vol I, 349-371, (C.P. Tsokos and I.N. Shimi, eds.), Academic Press, New York.

Block, H.W. and Basu, A.P. (1974) A continuous bivariate exponential extension, *J. Amer. Statist. Assoc.*, 69, 1031-1037.

Block, H.W. and Savits, T.H. (1977) *Multivariate IFRA Distributions*, Dept. of Math. and Statist., Research Report #77-04, University of Pittsburg, Pittsburgh, Pa.

Boas, Jr, R.P. (1954) *Entire Functions*. Academic Press, New York.

Bondesson, L. (1973a) Characterizations of the gamma distribution, *Teor. Veroyatnost. i Primenen*, 18, 382-384.

Bondesson, L. (1973b) Characterizations of the normal and the gamma distributions, *Z. Wahrscheinlichkeitstheorie verw. Geb.*, 26, 335-344.

Bondesson, L. (1974) Characterizations of probability laws through constant regression, *Z. Wahrscheinlichkeitstheorie verw. Geb.* 30, 93-115.

Bondesson, L. (1975) Characterizations of the gamma distribution and related laws, *Statistical Distributions in Scientific Work*, Vol. 3, 185-199 (G.P. Patil, S. Kotz and J.K. Ord. eds.) Dordrecht-Boston; D. Reidel.

Bondesson, L. (1977) *A note on sufficiency and independence*, Lund Univ., Dept. of Math. Statist., NFMS - 3046.

Bolger, E.T. and Harkness, W.L. (1965) Characterizations of some distributions by conditional moments, *Ann. Math. Statist.*, 36, 703-705.

Bochner, S. (1960) *Harmonic Analysis and the Theory of Probability*, University of California Press, Berkeley, Calif.

Brindley, Jr. E.C. and Thompson, Jr. W.A. (1972) Dependence and aging aspects of multivariate survival, *J. Amer. Statist. Assoc.*, 67, 822-830.

Buchanan, W.B. and Singpurwalla, N.D. (1977) Some stochastic characterizations of multivariate survival, *Theory and Applications of Reliability*, Vol. 2, 329-348 (C.P. Tsokos and I.N. Shimi, eds.), Academic Press, New York.

Cauchy, A.L. (1821) *Cours d'analyse de l'École Polytechnique*, Vol. I, Analyse algébrique, V. Paris, (*Œvres*, Ser. 2, Vol. 3, 98-113 and 220, Paris, 1897).

Chan, L.K. (1967) On a characterization of distributions by expected values of extreme order statistics, *Amer. Math. Monthly*, 74, 950-951.

Chong, K.M. (1977) On characterizations of the exponential and geometric distributions by expectations, *J. Amer. Statist. Assoc.*, 72, 160-161.

Chung, K.L. (1972) The Poisson process as a renewal process, *Periodica Mathematica Hungarica*, 2, 41-48.

Cifarelli, D.M. and Regazzini, E. (1974-76) Sulla caratterizzazione di una famiglia di distribuzioni basata sulla efficienza di una funzione campionaria, *Rend. Sem. Mat. Univ. e Politec. Torino*, 33 (1974/75), 299-311 (1976) (English summary).

Cinlar, E. and Jagers, P. (1973) Two mean values which characterize the Poisson process, *J. Appl. Prob.*, 10, 678-681.

Cox, D.R. (1962) *Renewal Theory*, Methuen, London.

Crawford, C.B. (1966) Characterization of geometric and exponential distributions, *Ann. Math. Statist.*, 37, 1790-1795.

Csörgö, M. and Seshadri, V. (1970) On the problem of replacing composite hypotheses by equivalent simple ones, *Rev. Int. Statist. Inst.*, 38, 351-368.

Csörgö, M. and Seshadri, V. (1971) Characterizing the Gaussian and exponential law via mappings onto the unit interval, *Z. Wahrscheinlichkeitstheorie verw. Geb.* 18, 333-339.

Csörgö, M., Seshadri, V. and Yalovsky, M. (1974) *Applications of characterizations in the area of goodness of fit*, Carleton Math. Series #112, Carleton University, Ottawa, Canada.

Csörgö, M., Seshadri, V. and Yalovsky, M. (1975) Applications of characterizations in the area of goodness of fit, *Statistical Distributions in Scientific Work*, Vol. 3, 79-90 (G.P. Patil, S. Kotz, J. Ord, eds.), Dordrecht-Boston; D. Reidel

Cundy, H. (1966) Birds and atoms, *Math. Gazette*, 50, 294-295.

Daley, D.J. (1973) Various concepts of orderliness for point processes, *Stochastic Geometry and Stochastic Analysis* (D.G. Kendall and E.F. Harding, eds.) Wiley, New York.

Daley, D.J. and Vere-Jones, D. (1972) A summary of the theory of point processes, *Stochastic Point Processes: Statistical Analysis, Theory and Applications*, 299-383 (P.A.W. Lewis, ed.) Wiley, New York.

Dallas, A.C. (1973) A characterization of the exponential distribution, *Bull. Soc. Math. Grece (N.S.)* 14, 172-175.

Dallas, A.C. (1975) On a characterization by conditional variance, (manuscript), Athens University, Greece.

Dallas, A.C. (1976) Characterizing the Pareto and power distributions, *Ann. Inst. Statist. Math.* 28A, 491-497.

Darboux, G. (1875) Sur la composition des forces en statique, *Bull. Sci. Math.*, 9, 281-288.

David, H.A. (1970) On Chiang's proportionality assumption in the theory of competing risks, *Biometrics*, 26, 336-339.

Davis, D.J. (1952) An analysis of some failure data, *J. Amer. Statist. Assoc.*, 47, 113-150.

Day, N.E. (1969) Linear and quadratic discrimination in pattern recognition, *IEEE Trans. on Inform. Theory*, 15, 419-421.

de Bruijn, N.G. (1966) On almost additive functions, *Colloqu. Math.*, 15, 59-63.

Deryagin, Yu.V., Polesizkaya, N.A. (1975) A characterization of the exponential distribution by the property of absence of an aftereffect, *Tr. Mosc. In-ta Elect. Mash.* 44, 192-198 (in Russian). (Analogous to *Sahobov and Geshev's* (1974) result, obtained independently.)

Desu, M.M. (1971) A characterization of the exponential distribution by order statistics, *Ann. Math. Statist.*, 42, 837-838.

Durrett, R.T. and Ghurye, S.C. (1976) Waiting times without memory, *J. Appl. Prob.* 13, 65-75.

Epstein, B. and Sobel, M. (1953) Life testing, *J. Amer. Statist. Assoc.*, 48, 486-502.

Epstein, B. and Sobel, M. (1954) Some theorems relevant to life testing from an exponential distribution, *Ann. Math. Statist.*, 25, 373-381.

Erickson, K.B. and Guess, H. (1973) A characterization of the exponential law, *Ann. Prob.*, 1, 183-185.

Esary, J.D. and Marshall, A.W. (1974) Multivariate distributions with exponential minimums, *Ann. Statist.*, 2, 84-93.

Esary, J.D. and Marshall, A.W. (1976) *Multivariate IHRA distributions* (unpublished report) (July).

Esary, J.D., Marshall, A.W. and Proschan, F. (1973) Shock models and wear processes, *Ann. Prob.*, 1, 627-650.

Farlie, D.J.G. (1960) The performance of some correlation coefficients for a general bivariate distribution, *Biometrika*, 47, 307-323.

Feller, W. (1966) *An Introduction to Probability Theory and its Applications*, Vol. II, Second Ed., J. Wiley, New York.

Ferguson, T.S. (1964) A characterization of the exponential distribution, *Ann. Math. Statist.*, 35, 1199-1207.

Ferguson, T.S. (1965) A characterization of the geometric distribution, *Amer. Math. Monthly*, 72, 256-260.

Ferguson, T.S. (1967) On characterizing distributions by properties of order statistics, *Sankhyā, Ser. A*, 29, 265-278.

Fisz, M. (1958) Characterization of some probability distributions, *Skand. Aktuarietidskr.*, 1-2, 65-70.

Fisz, M. (1963) *Probability Theory and Mathematical Statistics*, Third Ed., J. Wiley, New York.

Flusser, P. (1971) A property of the uniform distribution on compact Abelian groups with application to characterization problems in probability, *Accad. Naz. Dei. Lincei, Ser. 8*, 1, 151-155.

Fortet, R. (1977) *Elements of Probability Theory*, Gordon and Breach, New York.

Funk, G.M. (1972) *A characterization of probability distributions by conditional expectation*, (manuscript) Statistics Unit, Oklahoma State University, Stillwater, Oklahoma.

Fisher, R.A. and Tippett, L.H.C. (1928) Limiting forms of the frequency distributions of the largest or the smallest member of a sample, *Proc. Cambridge Philos. Soc.*, 24, 180-190.

Fréchet, M. (1927) Sur la loi de probabilité de l'ecart maximum, *Ann. Soc. Polon. Math. Crocovie*, 6, 93-116.

Fréchet, M. (1951) Sur les tableaux de corrélation dont les marges sont données, *Ann. Univ. Lion, A, Ser. 3*, 14, 53-77.

Friday, D.S. and Patil, G.P. (1977) A bivariate exponential model with applications to reliability and computer generation of random variables, *Theory and Applications of Reliability*, Vol. 1, 527-549 (C.P. Tsokos and I.N. Shimi, eds.) Academic Press, New York.

Freund, J.E. (1961) A bivariate extension of the exponential distribution, *J. Amer. Statist. Assoc.*, 56, 971-977.

Gajjar, A.V. and Khatri, C.G. (1969) A trivariate extension of Freund's bivariate distribution, *Vidya*, 12, 126-244.

Galambos, J. (1972) Characterization of certain populations by independence of order statistics, *J. Appl. Prob.*, 9, 224-230.

Galambos, J. (1975a) Characterizations of probability distributions by properties of order statistics, I, *Statistical Distributions in Scientific Work*, Vol. 3, 71-88, Dordrecht-Boston; D. Reidel.

Galambos, J. (1975b) Characterizations of probability distributions by properties of order statistics, II, *Statistical Distributions in Scientific Work*, Vol. 3, 89-101, Dordrecht-Boston; D. Reidel.

Galambos, J. (1975c) Characterizations in terms of properties of the smaller of two observations, *Commun. in Statist.*, 4, 239-244.

Galambos, J. (1978) *The Asymptotic Theory of Extreme Order Statistics*, J. Wiley, New York.

Garagorry, F.L. and Ahsanullah, M. (1977) A characterization of stationary renewal processes and of memoryless distributions, *Statist. Hefte (N.F.)*, 18, No. 1, 46-48.

Geary, R.C. (1936) Distribution of Student's ratio for non-normal samples, *J. Roy. Statist. Soc., Ser. B*, 3, 178-184.

Ghurye, S.C. (1960) Characterization of some location and scale parameter families of distributions, *Contrib. Prob. Statist.*, 202-215, Stanford University Press, Stanford, California.

Gokhale, D.V. (1975) Maximum entropy characterizations of some distributions, *Statistical Distributions in Scientific Work*, Vol. 3, 299-305, Dordrecht-Boston; D. Reidel.

Govindarajulu, Z. (1966a) Characterization of the exponential and power distribution, *Skand. Aktuarietidskr.*, 49, 132-136. (Errata sheet available from the author.)

Govindarajulu, Z. (1966b) Characterization of normal and generalized truncated normal distributions using order statistics, *Ann. Math. Statist.*, 37, 1011-1015.

Govindarajulu, Z. (1975) Characterization of the exponential distribution using lower moments of order statistics, *Statistical Distributions in Scientific Work*, Vol. 3, 117-129, Dordrecht-Boston; D. Reidel. (Errata sheet available from the author.)

Govindarajulu, Z., Huang, J.S. and Saleh, A.K.Md.E. (1975) Characterization of distributions using expected spacings between consecutive order statistics, *Statistical Distributions in Scientific Work*, Vol. 3, 143-147, Dordrecht-Boston; D. Reidel.

Goodman, I.R. and Kotz, S. (1973) On a θ-generalized multivariate distribution, *J. Multiv. Anal.*, 3, 204-219.

Gumbel, E.J. (1960) Bivariate exponential distributions, *J. Amer. Statist. Assoc.*, 55, 698-707.

Guerrieri, G. (1965) Some characteristic properties of the exponential distributions, *G. Economisti*, 24, 427-437 (in Italian).

Gupta, R.C. (1973) A characteristic property of the exponential distribution, *Sankhyā, B*, 35, 365-366.

Gupta, R.C. (1974) Characterization of distributions by a property of discrete order statistics, *Commun. in Statist.*, 3, 287-289.

Gupta, R.C. (1975) On characterization of distributions by conditional expectations, *Commun. in Statist.*, 4, 99-103.

Gupta, R.C. (1976) Some characterizations of distributions by properties of their forward and backward recurrence times in a renewal process, *Scand. J. of Statist.*, 3, 215-216.

Haines, A.M. and Singpurwalla, N.D. (1974) Some contributions to the stochastic characterization of wear, *Reliability and Biometry*, 47-80, SIAM, Philadelphia.

Hagstroem, K.G. (1960) Remarks on Pareto distributions, *Skand. Aktuarietidskr.*, 59-71.

Hamdan, M.A. (1972) On a characterization by conditional expectation, *Technometrics*, 14, 497-499.

Hamel, G. (1905) Eine Basis aller Zahlen und die unstetigen Lösungen der Funktionalgleichung f(x+y) = f(x) + f(y), *Math. Ann.*, 60, 459-462.

Harris, R. (1970) A multivariate definition for increasing hazard rate distribution functions, *Ann. Math. Statist.*, 41, 713-717.

Hawkins, D. and Kotz, S. (1976) On a clocking property of the exponential distribution, *Austr. J. of Statist.*, 18, 170-172.

Hoang, H.N. (1968) Estimation of stability of a characterization of an exponential distribution, *Litovs. Mat. Sbornik.*, 8, 175-177 (in Russian).

Hoeffding, W. (1953) On the distribution of the expected values of the order statistics, *Ann. Math. Statist.*, 24, 93-100.

Holmes, P.T. (1974) Another characterization of the Poisson process, *Sankhyā, A*, 36, 449-450.

Hogg, R.V. (1964) Applications of the characterizations of distributions to tests of fit, randomness and independence (Abstract), *Ann. Math. Statist.*, 35, 1837.

Huang, J.S. (1973) *Expected value of the spacings between order statistics* (manuscript) University of Guelph, Canada.

Huang, J.S. (1974) Characterizations of the exponential distribution by order statistics, *J. Appl. Prob.*, 11, 605-608.

Huang, J.S. (1974a) On a theorem of Ahsanullah and Rahman, *J. Appl. Prob.*, 11, 216-218.

Huang, J.S. (1974b) Characterizations of the exponential distribution by order statistics, *J. Appl. Prob.*, 11, 605-608.

Huang, J.S. (1975) A note on order statistics from Pareto distribution, *Skand. Aktuarietidskr.*, 3, 187-190.

Huang, J.S. (1975a) Characterization of distributions by the expected values of the order statistics, *Ann. Inst. Statist. Math.*, 27, 87-93.

Huang, J.S., Arnold, B.C. and Ghosh, M. (1977) On characterizations of the uniform distribution based on identically distributed spacings (to appear).

Isham, V., Shanbhag, D.N. and Westcott, M. (1975) A characterization of the Poisson process using forward recurrence times, *Math. Proc. Camb. Phil. Soc.*, 78, 515-516.

Johnson, N.L. and Kotz, S. (1970) *Continuous Univariate Distributions I*, J. Wiley, New York.

Johnson, N.L. and Kotz, S. (1972) *Distributions in Statistics: Continuous Multivariate Distributions*, J. Wiley, New York.

Johnson, N.L. and Kotz, S. (1973a) A vector valued multivariate hazard rate, I, Institute of Statistics *Mimeo Series No. 873*, Univ. of North Carolina, Chapel Hill, N.C.

Johnson, N.L. and Kotz, S. (1973b) A vector valued multivariate hazard rate, II, Institute of Statistics *Mimeo Series No. 883*, Univ. of North Carolina, Chapel Hill, N.C.

Johnson, N.L. and Kotz, S. (1975) A vector-valued multivariate hazard rate, *J. Multiv. Anal.*, 5, 53-66, Addendum *ibid.* 498.

Kac, M. (1939) On a characterization of the normal distribution, *Amer. J. Math.*, 61, 726-728.

Kadane, J.B. (1971) A moment problem for order statistics, *Ann. Math. Statist.*, 42, 745-751.

Kadane, J.B. (1974) A characterization of triangular arrays which are expectations of order statistics, *J. Appl. Prob.*, 11, 413-416.

Jurkat, W.B. (1965) On Cauchy's functional equation, *Proc. Amer. Math. Soc.*, 16, 683-686.

Kagan, A.M., Linnik, Yu.V. and Rao, C.R. (1973) *Characterization Problems in Mathematical Statistics*, J. Wiley, New York (English translation).

Kaminsky, K.S. and Nelson, P.I. (1975) Characterization of distributions by the form of predictors of order statistics, *Statistical Distributions in Scientific Work*, Vol. 3, 113-115, Dordrecht-Boston; D. Reidel.

Kawata, T. and Sakamoto, H. (1949) On the characterization of the normal population by the independence of the sample mean and the sample variance, *J. Math. Soc. Japan*, 1, 111-115.

Kelker, D. and Matthes, T.K. (1970) A sufficient statistics characterization of the normal distribution, *Ann. Math. Statist.*, 41, 1086-1090.

Kemperman, J.H.B. (1971) Product measures for which deviation and minimum are independent, *Sankhyā, Ser. A*, 33, 271-288.

Klebanov, L.B. (1977) Some results related to characterization of the exponential distributions, (Pre-print), Leningrad University, Leningrad, U.S.S.R. (in Russian).

Klebanov, L.B. and Melamed, I.A. (1976) The characterization of the normal and gamma distributions by properties of their Bayesian estimates, *Litovsk. Mat. Sb.*, 16, no. 1, 123-137, 247 (in Russian).

Kondo, T. (1931) Theory of sampling distribution of standard deviations, *Biometrika*, 22, 31-64 (especially pp. 52-53).

Konheim, A.G. (1971) A note on order statistics, *Amer. Math. Monthly*, 78, 524.

Kotlarski, I.I. (1972) On a characterization of some probability distributions by conditional expectation, *Sankhyā, Ser. A*, 34, 461-467.

Kotlarski, I.I. and Hinds, W.E. (1975) Characterization theorems using conditional expectations, *J. Appl. Prob.*, 12, 400-406.

Kotz, S. (1973) Normality vs. lognormality with applications, *Commun. in Statist.*, 1, 198-218.

Kotz, S. (1974) Characterizations of statistical distribution: a supplement to recent surveys, *Rev. Inst. Int. Statist.*, 42, 39-65.

Kotz, S. and Johnson, N.L. (1974) A characterization of exponential distributions by waiting time property, *Commun. in Statist.*, 3(3), 257-258.

Kovalenko, I.N. (1965) On a class of limit distributions for rarefied streams of events, *Litovsk. Mat. Sbornik*, 5, 569-573 (in Russian).

Krishnaji, N. (1970) Characterization of the Pareto distribution through a model of under-reported incomes, *Econometrica*, 38, 251-255.

Krishnaji, N. (1971) Note on a characterizing property of the exponential distribution, *Ann. Math. Statist.*, 42, 361-362.

Langberg, N., Proschan, F. and Quinzi, A.J. (1977) Transformations yielding reliability models based on independent random variables: a survey, *Applications of Statistics*, 323-337 (P.R. Krishnaiah, ed.) North Holland Publishing Co.

Laurent, A.G. (1972) On characterization of some distributions by truncation properties, *Res. Mem. No. S.C.A.G L. 15-0-2ii*, Wayne State Univ., Detroit, Mich. (also *Bull. of IMS*, 5, October 1972, Abstract 72t-59).

Laurent, A.G. (1974) On characterization of some distributions by truncation properties, *J. Amer. Statist. Assoc.*, 69, 823-827.

Laurent, A.G. and Gupta, R.C. (1969) A characterization of the exponential distribution (abstract), *Ann. Math. Statist.*, 40, 1865.

Lee, L. (1977) *Multivariate distributions having Weibull properties* (Pre-print), Dept. of Statistics, Virginia Polytechnic Institute, Blacksburg, Va.

Ling, K.D. (1968/69) Distributions of [kX] and [X+λ] of a continuous random variable X and its application in characterization problems, *Nanta Mathematica*, 3, 116-138.

Lobacevskii, N.L. (1829) On the foundations of geometry, III, §12, *Kasaner Bote*, 27, 227-243 (in Russian).

Loève, M. (1963) *Probability Theory*, 3rd ed., Van Nostrand, New York.

Lukacs, E. (1942) A characterization of the normal distribution, *Ann. Math. Statist.*, 13, 91-93.

Lukacs, E. (1977) Stability theorems, *Adv. Appl. Prob.*, 9, 336-361.

Lukacs, E. and Laha, R.G. (1964) *Applications of Characteristic Functions*, Griffin, London.

Mallows, C.L. (1973) Bounds on distributions functions in terms of expectations of order-statistics, *Ann. Prob.*, 1, 297-303.

Malmquist, S. (1950) On a property of order statistics from a rectangular distribution, *Skand Aktuarietidskr.*, 33, 214-222.

Marsaglia, G. and Tubilla, A. (1975) A note on the lack of memory property of the exponential distribution, *Ann. Prob.*, 3, 352-354.

Mardia, K.V. (1962) Multivariate Pareto distributions, *Ann. Math. Statist.*, 33, 1008-1015.

Marshall, A.W. (1975) Some comments on the hazard gradient, *Stoch. Proc. Appl.*, 3, 293-300.

Marshall, A.W. (1975a) Multivariate distributions with monotone hazard rate, *Reliability and Fault Tree Analysis*, 259-284 (R. Barlow, J. Fussel, N. Singpurwalla, eds.) SIAM, Philadelphia.

Marshall, A.W. and Olkin, I. (1967a) A multivariate exponential distribution, *J. Amer. Statist. Assoc.*, 62, 30-44.

Marshall, A.W. and Olkin, I. (1967b) A generalized bivariate exponential distribution, *J. Appl. Prob.*, 4, 291-302.

Mathai, A.M. (1967) On the structural properties of the conditional distributions, *Can. Math. Bull.*, 10, 239-245.

Menon, M.V. (1966) Characterization theorems for some univariate probability distributions, *J. Roy. Statist. Soc., Ser. B*, 28, 143-145.

Menon, M.V. and Seshadri, V. (1970) *A characterization theorem useful in hypothesis testing*, (manuscript), McGill University, Montreal, Canada.

Moeschberger, M.L. (1974) Life test under competing causes of failure, *Technometrics*, 16, 39-47.

Mogyorodi, J. (1969) A rarefaction of renewal processes, *Magyar Tud. Akad. III, Oszt. Kozl.*, 19, 25-31 (in Hungarian).

Mogyorodi, J. (1971) Some remarks on rarefaction of the renewal process, *Litovsk. Mat. Sborn.*, 11, 303-305.

Mogyorodi, J. (1972) On the rarefaction of renewal processes, I-II; III-IV, *Stud. Sci. Math. Hung.*, 7, 285-305; *ibid.* 8, 21-29.

Morgenstern, D. (1956) Einfache Beispiele zweidimensionaler Verteilungen, *Mitt. für Math. Stat.*, 8, 234-235.

Nabeya, S. (1950) On a relation between exponential law and Poisson's law, *Ann. Inst. Statist. Math.*, 2, 13-16.

Nagaraja, H.N. (1975) Characterization of some distributions by conditional moments, *J. Indian Statist. Assoc.*, 13, 57-61.

Nagaraja, H.N. (1977) On a characterization based on record values, *Austr. J. of Statist.*, 19, 71-74.

Obretenov, A. (1970) On a property of exponential distribution, *Fiz. Mat. Spisanie*, 13, 51-53 (in Bulgarian).

Oliker, V.I. and Singh, J. (1977) On a characterization of the Poisson distribution based on a damage model (Pre-print), Temple University, Philadephia, Pa.

Patil, G.P. and Ratnaparkhi, M.V. (1975) Problems of damaged random variables and related characterizations, *Statistical Distributions in Scientific Work*, Vol. 3, 255-270, Dordrecht-Boston; D. Reidel.

Patil, G.P. and Seshadri, V. (1964) Characterization theorems for some univariate probability distributions, *J. Roy. Statist. Soc., Ser. B*, 26, 286-292.

Paulson, A.S. (1973) A characterization of the exponential distribution and a bivariate exponential distribution, *Sankhyā, Ser. A*, 35, 69-78.

Paulson, A.S. and Uppuluri, V.R.R. (1972) A characterization of the geometric distribution and a bivariate geometric distribution, *Sankhyā, Ser. A*, 34, 88-91.

Pickands, J. III (1976) A class of multivariate negative exponential distributions (Pre-print), Dept. of Statistics, University of Pennsylvania, Philadelphia, Pa.

Pisarewska, H. (1975) Characterization of normal and exponential distributions, *Scien. Bull. Lódz Techn. Univ. Math. Series*, 7, No. 209, 59-65.

Pollak, M. (1973) On equal distributions, *Ann. Statist.*, 1, 180-182.

Polya, G. (1923) Verteilung des Gaussschen Fählergesetzes aus einer Funktionalgleichung, *Math. Z.*, 18, 96-108.

Proschan, F. and Sullo, P. (1976) Estimating the parameters of a multivariate exponential distribution, *J. Amer. Statist. Assoc.*, 71, 465-472.

Puri, P.S. (1970) Characterizations of distributions via order statistics (Abstract), *Ann. Math. Statist.*, 41, 1154.

Puri, P.S. (1972) On a property of exponential and geometric distributions (Abstract 72t-53), *IMS Bull.*, 1.

Puri, P.S. (1973) On a property of exponential and geometric distributions and its relevance to multivate failure rate, *Sankhyā, Ser. A*, 35, 61-68.

Puri, P.S. and Rubin, H. (1970) A characterization based on the absolute difference of two i.i.d. random variables, *Ann. Math. Statist.*, 41, 2113-2122.

Puri, P.S. and Rubin, H. (1974) On a characterization of the family of distributions with constant multivariate failure rates, *Ann. Prob.*, 2, 738-740.

Rao, C.R. and Rubin, H. (1964) On a characterization of the Poisson distribution, *Sankhyā, Ser. A*, 26, 294-298.

Reinhardt, H.E. (1968) Characterizing the exponential distribution, *Biometrics*, 24, 437-438.

Revankar, N.S., Hartley, M.J. and Pagano, M. (1974) A characterization of the Pareto distribution, *Ann. Statist.*, 2, 599-601.

Rényi, A. (1953) On the theory of order statistics, *Acta Math., Acad. Sci. Hungar.*, 4, 191-232.

Rényi, A. (1956) A characterization of the Poisson process, *Magyar Tud. Akad. Mat. Kutato Int. Kozl.*, 1, 519-527 (in Hungarian). (Translated into English in *Selected Papers of Alfrêd Rényi*, Vol. 1, Akademiai Kiadó, Budapest, 1976).

Rényi, A. (1967) Remarks on the Poisson process, *Studia Sci. Math. Hungar.*, 2, 119-123.

Reza, F.M. (1961) *An Introduction to Information Theory*, McGraw-Hill, New York.

Rogers, G.S. (1959) A note on the stochastic independence of order statistics, *Ann. Math. Statist.*, 30, 1263-1264.

Rogers, G.S. (1963) An alternate proof of the characterization of the density Ax^B, *Amer. Math. Monthly*, 70, 857-858.

Rossberg, H.J. (1960) Über die Verteilungsfunktionen der Differenzen und Quotienten von Ranggrössen, *Math. Nachr.*, 21, 37-79.

Rossberg, H.J. (1964) Über die Eigenschaften gewisser Funktionen von Ranggrössen, *Abhandl. Deutsch. Akad. Wiss. Berlin*, 4, 103-105.

Rossberg, H.J. (1965) Über die stochastische Unabhängigkeit gewisser Funktionen von Ranggrössen, *Math. Nachr.*, 28 , 157-167.

Rossberg, H.J. (1966) Characteisierungsprobleme, die sich aus der von A. Rényi in die Theorie der Ranggrössen eingeführten Methode ergeben, *Monatsberichte Deutche Akad., Wiss., Berlin*, 8, 561-572.

Rossberg, H.J. (1968) Eigenschaften der charakteristischen Funktionen von einseitig beschrankten Verteilungsfunktionen und ihre Anwendung auf ein Charackerisierungsprobleme der mathematischen Statistik, *Math. Nachr.*, 37, 37-57.

Rossberg, H.J. (1972a) Characterization of the exponential and the Pareto distributions by means of some properties of the distributions which the differences and quotients of order statistics are subject to, *Math. Operationsforschung und Statistik*, 3, 207-216.

Rossberg, H.J. (1972b) Characterization of distribution functions by the independence of certain functions of order statistics, *Sankhyā, Ser. A*, 34, 111-120.

Sagrista, S.N. (1952) On a generalization of Pearson's curves to the two-dimensional case, *Trab. de Estadist.*, 3, 273-314.

Sahobov, O.M. and Geshev, A.A. (1974) Characteristic properties of the exponential distribution, *Natura. Univ. Plovdiv.*, 7, No. 1, 25-28 (in Russian). (Cf. *Deryagin, et al.* (1975).)

Samanta, M. (1972) Characterization of the Pareto distribution, *Skand. Aktuarietidskr.*, 55, 191-192.

Seshadri, V., Csörgö, M. and Stephens, M.A. (1969) Tests for the exponential distribution using Kolmogorov-type statistics, *J. Roy. Statist. Soc., Ser. B*, 31, 499-509.

Seshadri, V. and Patil, G.P. (1964) A characterization of bivariate distribution by the marginal and the conditional distribution of the same component, *Ann. Inst. Statist. Math.*, 15, 215-221.

Serfozo, R.F. (1977) Compositions, inverses and thinnings of random measures, *Z. Wahrscheinlichketistheorie verw. Geb.*, 37, 253-265.

Sethuraman, J. (1965) On a characterization of the three limiting types of extremes, *Sankhyā, Ser. A*, 27, 357-364.

Sethuraman, J. (1976) Characterization of distributions by large deviation rates, (Pre-print), Florida State University, Tallahassee, Fl.

Shanbhag, D.N. (1970) Characterizations for exponential and geometric distributions, *J. Amer. Statist. Assoc.*, 65, 1256-1259.

Shanbhag, D.N. (1974) An elementary proof of the Rao-Rubin characterization of the Poisson distribution, *J. Appl. Prob.*, 11, 211-215.

Shanbhag, D.N. (1977) An extension of the Rao-Rubin characterization, *J. Appl. Prob.*, 14, 640-646.

Shanbhag, D.N. and Bhaskara Rao, M. (1975) A note on characterizations of probability distributions based on conditional expected values, *Sankhyā, Ser. A*, 37, 297-300.

Shanbhag, D.N. and Panaretos, J. (1977) Some results related to the Rao-Rubin characterization of the Poisson distribution, to appear in *Austr. J. of Statist.*

Shantaram, R. and Harkness, W. (1972) On a certain class of limit distributions, *Ann. Math. Statist.*, 43, 2067-2071.

Shohat, J.A. and Tamarkin, J.D. (1943) *The Problem of Moments*, Mathematical Surveys No. 1, American Math. Society, Providence, R.I.

Srivastava, M.S. (1967) A characterization of the exponential distribution, *Amer. Math. Monthly*, 74, 414-416.

Srivastava, R.C. (1971) On a characterization of the Poisson process, *J. Appl. Prob.*, 8, 615-616.

Srivastava, R.C. (1974) Two characterizations of the geometric distribution, *J. Amer. Statist. Assoc.*, 69, 267-269.

Srivastava, R.C. and Singh, J. (1975) On some characterizations of the binomial and Poisson distributions based on a damage model, *Statistical Distributions in Scientific Work*, Vol. 3, 271-277, Dordrecht-Boston; D. Riedel.

Srivastava, R.C. and Srivastava, A.B.L. (1970) On a characterization of the Poisson distribution, *J. Appl. Prob.*, 7, 495-501.

Srivastava, R.C. and Wang, Y.H. (1972) On characterization of certain populations by order statistics, *IMS Bull.*, 1, Abstract 136-24.

Srivastava, R.C. and Wang, Y.H. (1971) On characterization of the exponential and related distributions, *Ann. Math. Statist.*, 42, Abstract 1479.

Srivastava, R.C. and Wang, Y.H. (1978) *A characterization of the exponential and related distribution by linear regression* (manuscript), Ohio State University, Columbus, Ohio (to appear in *Ann. of Statist.*).

Steffensen, J.F. (1930) *Some Recent Research in the Theory of Statistics and Actuarial Science*, Cambridge University Press, Cambridge, U.K.

Steyn, H.S. (1960) On regression properties of multivariate probability functions of Pearson's type, *Proc. Roy. Acad. Sci.*, Amsterdam, 63, 302-311.

Sukhatme, P.V. (1937) Tests of significance for samples of the χ^2 population with two degrees of freedom, *Ann. of Eugenics*, 8, 52-56.

Swartz, B. (1975) A short proof of a characterization by conditional expectation, *IEEE Trans. Reliab.*, R-24, 76.

Szántai, R. (1971) On limiting distributions for the sums of random number of random variables concerning the rarefaction of recurrent processes, *Studia Sci. Math. Hungarica*, 6, 443-452.

Szász, D. (1976) Note on the literature concerning rarefactions, *Selected Papers of Alfréd Rényi*, Vol. 1, 279-280, Akad. Kiado, Budapest.

Szynal, D. (1976) On limit distribution theorems for sums of a random number of random variables appearing in the study of rarefaction of a recurrent process, *Appl. Math.*, 15, 3, 277-288.

Talwalker, S. (1970) A characterization of the double Poisson distribution, *Sankhyā, Ser. A*, 32, 265-270.

Talwalker, S. (1977) A note on characterization by the conditional expectation, *Metrika*, 24, 129-136.

Tanis, E.A. (1964) Linear forms in the order statistics from an exponential distribution, *Ann. Math. Statist.*, 35, 270-276.

Tata, M.N. (1969) On outstanding values in a sequence of random variables, *Z. Wahrscheinlichketistheorie verw. Geb.*, 12, 9-20.

Teicher, H. (1961) Maximum likelihood characterization of distributions, *Ann. Math. Statist.*, 32, 1214-1222.

Teissier, G. (1934) Recherches sur le vicillissement et sur les lois de mortalité *Ann. Phys. Biol. Phys.-Chem.*, 10, 237-264.

van Uven, M.T. (1947) Extension of Pearson's probability distributions to two variables, I-IV, *Proc. Roy. Acad. Sci.*, Amsterdam, 50, 1063-1070, 1252-1264.

Vartak, M.N. (1974) Characterization of certain classes of probability distributions, *J. Ind. Statist. Assoc.*, 12, 67-74.

Wang, P.C.C. (1975) Characterizations of the Poisson distribution based on random splitting and random expanding, *Disc. Math.*, 13, No. 1, 85-93.

Wang, Y.H. (1971) On characterization of some probability distributions and estimation of the parameters of the Pareto distribution, *Ph.D. Thesis*, Ohio State University, Columbus, Ohio.

Wang, Y.H. (1976) A functional equation and its application to the characterization of the Weibull and stable distributions, *J. Appl. Prob.*, 13, 385-391.

Wang, Y.H. and Chang, S.A. (1977) A new approach to the nonparametric tests of exponential distributions with unknown parameters, *The Theory and Applications of Reliability*, Vol. 2 (C. Tsokos and I.N. Shimi eds.) Academic Press, New York.

Weibull, W. (1939) The phenomenon of rupture in solids, *Ingen. Vetensk. Akad. Handl.*, 153.

Weibull, W. (1951) A statistical distribution function of wide applicability, *J. Appl. Mech.*, 18, 293-297.

Westcott, M. (1973) Some remarks on a property of the Poisson process, *Sankhyā, Ser. A.*, 35, 29-34.

Wilson, E.B. (1899) Note on the function satisfying the functional relation $\phi(u)\phi(v) = \phi(u+v)$, *Ann. of Math.*, 1, 47-48.

Young, G.S. (1958) The linear functional equation, *Amer. Math. Monthly*, 65, 37-38

Zolotarev, V.M. (1976) The stability phenomenon in the characterization of distributions, *Zapiski Nauch. Sem., Leningrad Otd. Mat. Inst. Steklov*, 61, 38-55 (in Russian).

SUBJECT INDEX

(For various types of characterizations of the exponential distribution and its monotone transforms the reader is referred to the *Table of Contents*.)

A

B

C

D

E

F

G

H

I

K

L

M

N

O

P

R

S

T

U

W